TECHNOLOGY IN THE 1990s: DEVELOPMENTS IN HYDRAULIC CEMENTS

TECHNOLOGY IN THE 1990s: DEVELOPMENTS IN HYDRAULIC CEMENTS

PROCEEDINGS OF
A ROYAL SOCIETY DISCUSSION MEETING
HELD ON 16 AND 17 FEBRUARY 1983

ORGANIZED AND EDITED BY
SIR PETER HIRSCH, F.R.S.,
J. D. BIRCHALL, F.R.S.,
D. D. DOUBLE, A. KELLY, F.R.S.,
G. K. MOIR AND C. D. POMEROY

LONDON
THE ROYAL SOCIETY
1983

Printed in Great Britain for the Royal Society
at the
University Press, Cambridge

ISBN 0 85403 215 0

First published in *Philosophical Transactions of the Royal Society of London*,
series A, volume 310 (no. 1511), pages 1–207

Published by the Royal Society
6 Carlton House Terrace, London SW1Y 5AG

CONTENTS

[Nine plates]

Glossary

In describing cement compounds a shortened notation is used in which the oxides are represented as:

$$C = CaO, \quad S = SiO_2, \quad A = Al_2O_3, \quad M = MgO,$$
$$F = Fe_2O_3, \quad \bar{S} = SO_3, \quad H = H_2O, \quad N = Na_2O, \quad K = K_2O.$$

Thus tricalcium silicate $3CaO.SiO_2 = C_3S$, tricalcium aluminate $3CaO.Al_2O_3 = C_3A$ and calcium hydroxide $Ca(OH)_2 = CH$, for example.

AFm phase	typically a tetracalcium monosulphoaluminate hydrate, usually with some substitution of Al by Fe
AFt phase	typically ettringite, usually with some substitution of Al by Fe
alite	impure form of C_3S ($3CaO.SiO_2$) found in commercial Portland cement clinker
alkali modified belite	$KC_{23}S_{12} = K_2O.23CaO.12SiO_2$
aluminate phase	$C_3A = 3CaO.Al_2O_3$
anhydrite	$C\bar{S} = CaSO_4$
a.r.	alumina ratio
belite	impure form of C_2S ($2CaO.SiO_2$) found in commercial Portland cement clinker
brucite	$MH = Mg(OH)_2$
calcium langbeinite	$C_2K\bar{S}_3 = 2CaSO_4.K_2SO_4$
calcium sulphosilicate	$C_5S_2\bar{S} = 4CaO.2SiO_2.CaSO_4$
cement clinker	nodular product from a cement manufacturing kiln
cement mortar	cement paste with sand filler
cement paste	mixture of cement powder with water
concrete	mixture of cement, sand and pebble aggregates
C–S–H gel	calcium silicate hydrate, a colloidal and mainly amorphous gel with a rather variable composition, which is the major hydration product of Portland cement
C/S ratio	molar ratio of CaO/SiO_2
D.S.P.	densified systems containing homogeneously arranged ultrafine particles
ettringite	$C_3A.3C\bar{S}.32H = 3CaO.Al_2O_3.3CaSO_4.32H_2O$
ferrite phase	$C_4AF = 4CaO.Al_2O_3.Fe_2O_3$
F.R.C.	fibre reinforced concrete or cement
free lime	$C = CaO$
G.R.C.	glass fibre reinforced cement
gypsum	$C\bar{S}.2H$ = calcium sulphate dihydrate $CaSO_4.2H_2O$
H.C.P.	hardened cement paste
l.o.p.	limit of proportionality
l.s.f.	lime saturation factor
MDF	macro-defect-free
monosulphoaluminate	$C_3A.C\bar{S}.12H = 3CaO.Al_2O_3.CaSO_4.12H_2O$
m.o.r.	modulus of rupture

oldhamite	calcium sulphide
O.P.C.	ordinary Portland cement
periclase	M = MgO
P.F.A.	pulverized fuel ash
P.I.C.	polymer impregnated concrete
portlandite	CH = crystalline calcium hydroxide $Ca(OH)_2$
q.X.r.d	quantitative X-ray diffractometry
R.H.P.C.	rapid hardening Portland cement
s.r.	silica ratio
u.t.s.	ultimate tensile strength
w/c ratio	ratio of mass of water to mass of cement in a cement paste
w/s ratio	ratio of mass of water to mass of solid in a paste mixture
X.p.s.	X-ray photoelectron spectrometry
X.r.d.	X-ray diffraction

Phil. Trans. R. Soc. Lond. A **310**, 5 (1983)
Printed in Great Britain

Introductory remarks

By Sir Peter Hirsch, F.R.S.
Department of Metallurgy and Science of Materials, University of Oxford, U.K.

World manufacture of hydraulic cements is close to one thousand million tonnes per year, and cements along with steel are the most important constructional materials. Yet our understanding of the structure and properties of cements is not anything like as advanced as that of steels. But the cheapness and ready availability of the primary product, the facts that the energy consumed in manufacture is considerably smaller than that for metals, plastics, or ceramics and that hardening takes place with water at ordinary temperatures, provides the motivation for improving the strength, toughness and durability of cements not only for their more conventional uses, but also so that they might be used in quite new applications as replacements for energy intensive plastics, metals and ceramics. During the last 5–10 years there have been significant advances in fundamental understanding of the physics and chemistry of cements, and various recipes are available for cements with much improved properties. The aims of this meeting are to review the present understanding of structure–property relations, and cement hydration, to describe the improvements in properties that have been obtained so far, and to discuss the potential applications of such materials. The organizers hope that this meeting, which has attracted people from different disciplines, researchers, and industrial producers and users, will not only help to stimulate research aimed at further improvements, but even more important, also help to promote the very difficult process of the transfer of the technological advances.

Phil. Trans. R. Soc. Lond. A **310**, 7–15 (1983)

Printed in Great Britain

Cement in its conventional uses: problems and possibilities

By W. A. Gutteridge and C. D. Pomeroy

Cement and Concrete Association, Wexham Springs, Slough SL3 6PL, U.K.

Although Portland cement is used almost exclusively in the construction industry it can nevertheless form the basis of certain special products that perform similarly to cast iron, timber or aluminium. In this paper, some of the parameters that characterize Portland cement are discussed in relation to the requirements of the user. It is acknowledged that as a consequence of improved manufacturing processes modern Portland cements are not identical with those produced several decades ago, but there are misconceptions about these changes, which are discussed.

Introduction

In the John Player lecture 'Concrete, an alternative material' (Pomeroy 1978), it was shown that Portland cement is potentially a versatile raw material that can be used in combination with other, usually more expensive, materials to manufacture products that have in the past been fabricated from plastics, timber or metals. Portland cement-based products are generally of low energy, particularly when assessed on the basis of performance, such as load bearing capacity (Pomeroy 1978). Table 1 provides some comparative figures that emphasize this point.

During the period when oil prices were low, materials research throughout the world was dominated by the organic chemist and metallurgist who had little interest in inorganic materials, particularly those having a direct relevance to the ceramic, building product and construction industries. More recently it has been widely realized that inorganic materials are more useful than was believed and in particular that hydraulic Portland cements have considerable potential for development and use in novel ways. The necessary materials are in abundant supply and available worldwide.

The principal reason why cement-based products have been confined to construction is that concrete is weak in tension and steel reinforcement must be provided to carry the tensile loads. Because steel rusts it must be protected and the high alkalinity of hydrated cement will provide this protection when it is sufficiently thick and impermeable. Thus traditionally, the use of concrete is largely confined to fairly massive elements that have adequate (20–75 mm) cover to the reinforcement, dependent upon the exposure environment and use of the element. Plain, unreinforced concrete is used in compression, building blocks for example.

In this introductory paper some of the factors that control the performance of Portland cements are discussed. Subsequent papers in this Seminar elaborate on some of the associated topics including ways in which cement-based materials can be tailored by the selective inclusion of fillers or polymers.

Portland cement manufacture

Portland cement clinker is made from a carefully proportioned and finely ground mixture of calcareous and argillaceous materials blended together either in the dry state or slurried

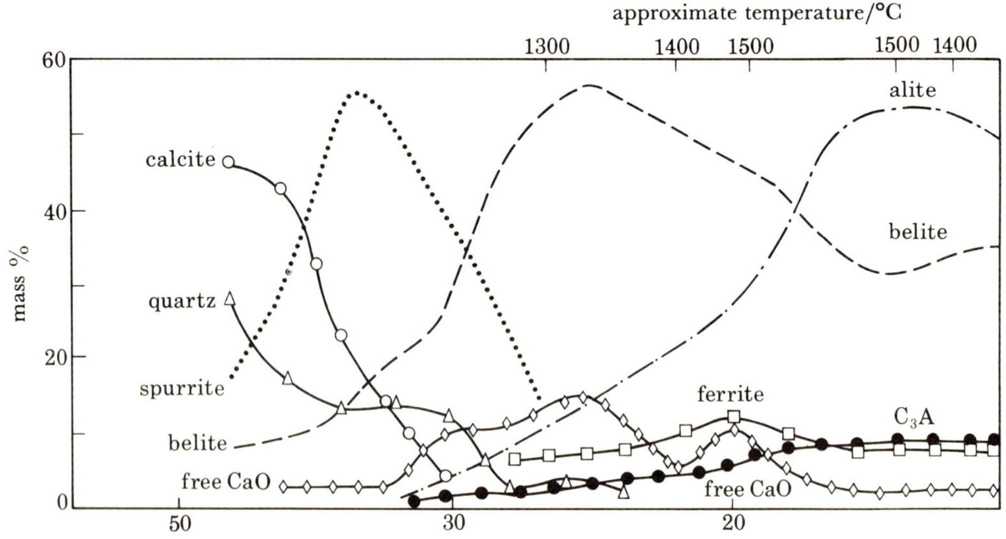

FIGURE 1. An example of the formation of the cement minerals within a kiln.

TABLE 1. TYPICAL MATERIAL PROPERTIES, COMPARED ON A PRODUCTION ENERGY BASIS

material	density / t/m	elastic modulus / GPa	tensile strength / MPa	energy per metre / GJ	(energy per metre / unit tensile strength) / kJ
aluminium	2.8	70	100	360	3.6
copper	8.9	130	200	530	2.6
mild steel	7.8	210	300	300	1.0
cast iron	7.9	150	150	360	2.4
glass	2.5	65	60	50	0.8
concrete	2.4	30	6	3.4	0.6
reinforced concrete	2.5	35	30	3.9	0.13

in water to produce a homogeneous raw feed of a specified composition (defined conventionally in terms of the oxides of calcium, aluminium, silicon and iron). This raw meal is burnt in a large rotary kiln (fired either by pulverized coal, oil or natural gas) and is fed gradually along the temperature gradient within the kiln to reach a maximum temperature of approximately 1500 °C. The raw mix is transformed by a series of reactions that take place between 400 and 1500 °C; this reaction sequence is illustrated in figure 1. Sintering and partial fusion occur and hard granular nodules of cement clinker form, which emerge from the lower end of the kiln into a cooling unit. In the cooling unit, heat exchange takes place between the clinker and combustion air, and the clinker exits at a temperature at which it can be conveyed and stored.

Experience has shown that reducting conditions during the burning and cooling processes should be avoided. Clinker that has experienced reduction will, in general, yield a cement with inferior strength properties to those prepared from oxidized clinker. Cement properties can also be influenced by the clinker cooling rate in the temperature interval 1500–1200 °C. Cooling through this interval is normally achieved in the kiln, and rapid cooling generally has a beneficial effect on cement strength properties. The clinker is ground in ball mills together with a small quantity of gypsum or other form of calcium sulphate to act as a retarder. Cement fineness, which is one of the important variables that will ultimately influence the compressive strength properties of the cement, is normally expressed in the form of surface area determined by an air permeability method. A typical fineness is 350 m^2 kg^{-1}, at which fineness approximately 90 % of the particles will be finer than 45 μm.

Within the framework of raw feed composition, firing temperature, rate of cooling and grinding, it is possible to produce Portland cements with different properties. Two common examples are ordinary Portland cement and sulphate resisting Portland cement.

Portland cement constituents

There has been a considerable amount of research performed on cement. In the late nineteenth century Le Chatelier (1882*a*, *b*) and Tornebohm (1897) had published their findings as to the composition of cement. It is interesting to recall that by 1898 there were probably 1000 bottle kilns on the Thames and Medway alone, producing some 25000 tons of cement clinker a week (Francis 1977). By the early 1930s there was substantial agreement as to the description of the principal phases, which Tornebohm had named alite, belite, celite, felite and an isotropic residue. Alite was generally accepted to be essentially tricalcium silicate. Tavasci (1934) had shown that polished samples of cement clinker, suitably etched and viewed under an optical microscope in reflecting light, could be used to study the clinker phases. Bogue (1929) formulated a potential compound composition and during the post war years Jeffrey (1952), C. M. Midgley (1952), H. G. Midgley (1957), Yamaguchi & Uchikawa (1961), Regourd (1967), Maki (1974) and many others have extended our knowledge of the phase composition of the cement clinker. Mills (1968) Richartz (1968) and Chatterji (1969) were among the early workers who used the electron microscope to examine the morphological features produced during cement hydration. The electron probe microanalyser has been used to study the minor compounds, or so-called impurities, which enter in solid solution in the principal phases. These minor compounds originate principally from the accessory minerals of the rocks from which the raw feed is made, from the fuel, refractory linings, worn parts of equipment and from admixtures or mineralizers, which may have been added to the raw feed.

Alite is now recognized as a substituted tricalcium silicate having a total of seven polymorphic forms whose structures have either a rhombohedral, monoclinic or triclinic unit cell. It is rare to find the triclinic variety in commercial U.K. clinker. Maki (1979) has reported that the monoclinic alite found within clinkers produced by modern processes (and containing MgO) can occur in either the M_1 or M_3 form or as a combination of these, depending upon the amount of MgO that has gone into solid solution. Thus the composition of an alite in a cement is far from a unique parameter; although the different forms may be expected to have different hydraulic properties, lattice defects are also important.

The term 'belite' is now reserved to describe the monoclinic or β form of the substituted dicalcium silicate that occurs in cement clinker. Other forms can occur, namely α, α′ and γ, but the latter rarely occurs in commercial U.K. clinker.

The terms 'celite' and 'felite' are rarely used now. The ferrite and the tricalcium aluminate phases are described as 'interstitial'. In a Portland cement the ferrite phase extends from approximately C_6AF_2 to C_6A_2F and is often substituted with several minor impurities. The tricalcium aluminate phase usually occurs in a form that has either a cubic, an orthorhombic or a tetragonal unit cell. In clinkers with an excess of alkalis over sulphate, the alkalis Na_2O and K_2O may be present in the C_3A solid solution and their presence has a distinct influence upon the reactivity of this phase. Periclase (MgO), free lime (CaO) and Apthitalite ($K_3Na(SO_4)_2$)

are some of the minor phases often present in clinker. The identification and quantitative analysis of a clinker by techniques such as Bogue potential compound composition, optical point counting and X-ray diffraction have become routine procedure, each associated with a particular precision. There is no procedure available at present that will produce an absolute quantitative analysis for the phases present. Numerous authors have reported results relating to the quantitative analyses obtained by the various procedures. At the Cement and Concrete Association both Bogue analysis and quantitative X-ray powder diffraction are used to describe a Portland cement. Results obtained from an X.r.d. analysis of a typical U.K. Portland cement are presented in figure 2.

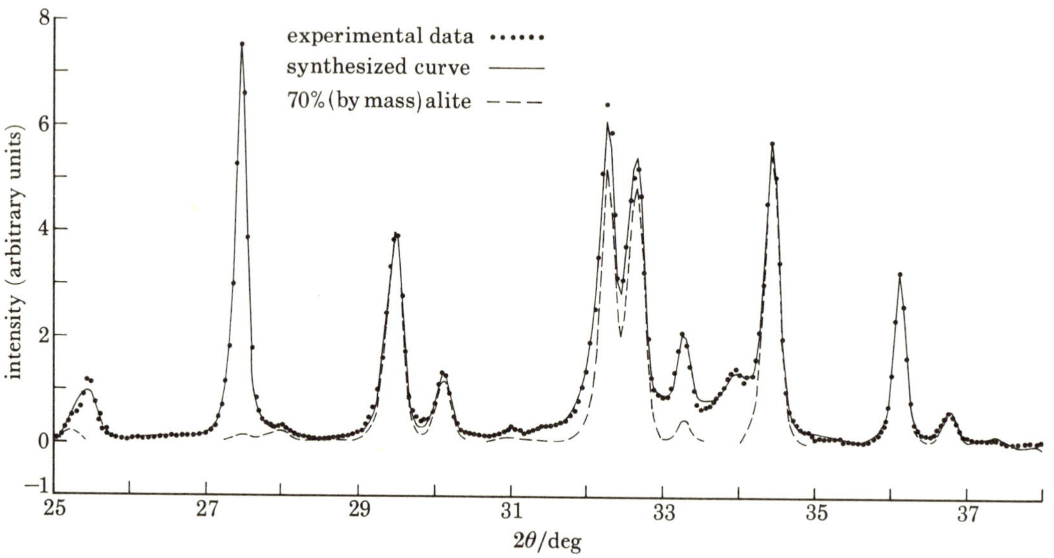

FIGURE 2. X.r.d. analysis of a typical ordinary Portland cement (2θ is the angle of deviation of the X-ray beam). The diffraction profiles for some components, used to obtain the analysis, are not shown: namely 11% belite, 6% $C_3A(0)$, 1% $C_3A(c)$, 6% ferrite, 4.5% $CaSO_4$, 0.5% CaO (all percentages by mass).

Developments in the performance of Portland cements

Views have been expressed that modern Portland cements are very different from those that were manufactured several decades ago. Corish & Jackson (1982) have published tables that show that in the U.K. the amount of alite in a typical cement has risen from about 14% in 1848 to 40% in 1939 and is in the range 45–64% today. This increase has usually been accompanied by a reduction in the C_2S. Many people believe that the changes in behaviour, which have occurred over the years, such as the rates of hydration and strength gain, can be attributed exclusively to the change in alite:C_2S ratio and to the greater finenesses of modern cements. While these factors are important, they provide only part of the answer and other parameters can be equally important. In fact, for U.K. cements, the data published by Corish & Jackson (1982) show no general increase in the fineness of cement produced in the years between 1960 and 1980, although the particle sizes and gradings of a particular cement may have changed over that period. Results of the BS 4550 concrete cube-crushing tests showed an average increase for 3 day (20 °C wet-cured) strengths from 16 to 24 N/mm^2 and for 28 day strengths from 35 to 44 N/mm^2 during the 20 year period.

It can thus be seen that there have been changes in cement manufacture that have altered cement performance. On the single basis of potential early strength, or even maximum 28 day strength, it can be claimed that improvements have been made in the quality of the cement, but concurrent with these changes is an increase in the reactivity, leading to higher heats of hydration and also to changes in the morphology or pore structure of the hardened cement paste. The requirements of cement users vary widely: the precaster may demand very high early strengths so that his production rate is maximized; the foundations engineer will require a low-heat cement to keep the temperature rise and gradients in mass concrete within acceptable bounds; other engineers may seek a cement that will have a high resistance to chemical attack or to the permeation of gases or liquids. Some of these and other objectives place conflicting requirements on the formulation of a particular cement.

TABLE 2. CEMENT PHASES

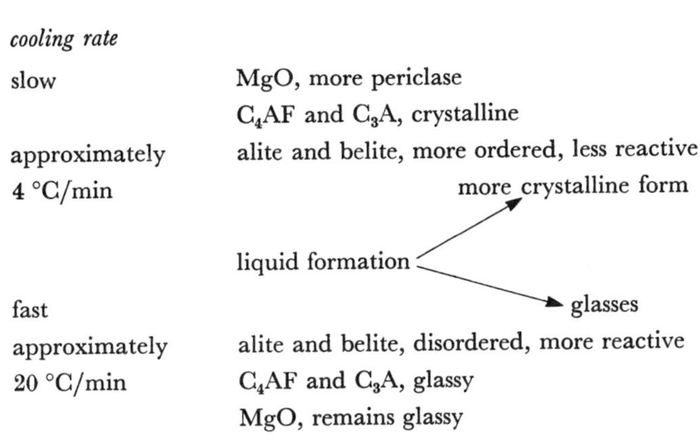

cooling rate	
slow	MgO, more periclase
	C_4AF and C_3A, crystalline
approximately	alite and belite, more ordered, less reactive
4 °C/min	more crystalline form
	liquid formation → more crystalline form / → glasses
fast	glasses
approximately	alite and belite, disordered, more reactive
20 °C/min	C_4AF and C_3A, glassy
	MgO, remains glassy

TABLE 3. CEMENT PROPERTIES

cooling rate	
	less hydraulically active
slow	larger particles (crystallites)
	higher early strength, lower 28 day strength
approximately	easier to grind
4 °C/min	less sound
	less sulphate resistant
	more sulphate resistant
fast	more sound
	harder to grind
approximately	lower early strength, higher 28 day strength
20 °C/min	smaller particles (crystallites)
	more hydraulically active

As mentioned before, the fineness to which the clinker has been ground is one of the factors influencing the compressive strength of a cement. Regourd *et al.* (1978) and Bernard (1978) have shown that hydration is more complete when a cement has been jet milled rather than ball milled. In general the reactivity and hence the strength at early age is increased by an increase in the content of C_3S, C_3A and soluble alkali in the clinker or by finer cement grinding.

The cooling rate of the clinker can affect the size and form of the cement phases (Chatterjee & Ghosh 1980), see table 2, these changes having an influence on the hydraulic activity of the cement (table 3).

The w/c ratio in a cement paste determines the amount of space in which hydration can occur and has an influence upon the final physical form of the hydration product. The degree of supersaturation within the paste fluid increases with decreasing w/c ratio and influences the rate at which nucleation can occur. Curing temperature has a pronounced effect upon hydration and not only affects the rate at which reaction occurs but also the hydration products

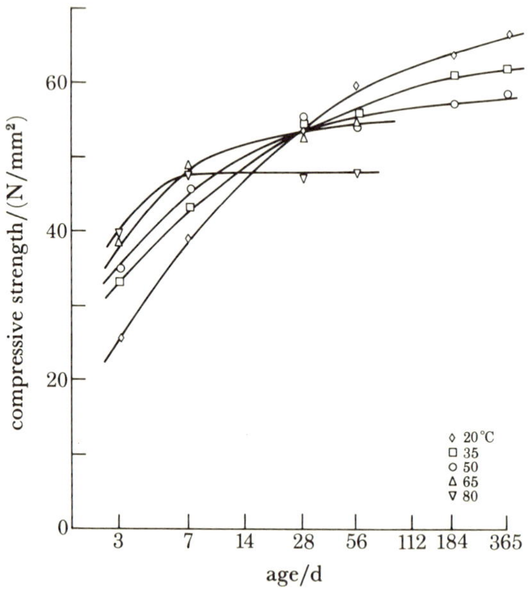

FIGURE 3. Effect of curing temperature upon the compressive strength of plain cement mortars.

(figure 3, Dalziel 1980). A higher curing temperature has a beneficial effect on strength at early ages but at later ages it may be detrimental. The pore sizes and pore volume within the cement paste are influenced by the amount of reaction induced by the temperature increase. It is convenient to divide the pore system into capillary pores and gel pores. Mercury porosimetry has shown that large capillary pores occur within the cement paste matrix between the unhydrated clinker particles and the hydration products. Total pore volume and the size of pores decrease as hydration proceeds and this reduces the permeability of the hardened cement paste. There are therefore several factors which affect the hydration and ultimately the strength attained by cement. Most hydraulic phases within the cement contribute to strength via reduction in porosity, which results from their hydration.

Concluding remarks

An attempt has been made to illustrate some of the ways in which cements differ and how these differences affect their hydration. In the past, a dominant objective has been to achieve strength development and close relationships were obtained between the total porosity of hydrated cement and compressive strength (Lawrence 1969; Roy & Gouda 1973). Recently, it has been realized that the pore size distribution is sensitive to the cement characteristics, the duration of hydration and the curing régime. Parrott (this symposium) shows clearly that such changes in distribution have a significant effect on the potential ingress of chemicals into the paste. There is thus a growing interest in the way that cement constitution relates to the pore structure within the hydrate, since an open hydrate structure will permit liquids or gases to enter more freely, which can result in degradation of the concrete, such as sulphate or chemical attack and depassivation of steel reinforcement consequent upon carbonation.

More rapidly reacting cements, and those containing P.F.A., ground granulated slag or a pozzolana, are all more sensitive to curing temperature and duration of exposure to a moist environment, and there is evidence that it is the pore structure that is most affected (see, for example, Scislewski 1975). Such cements are *less tolerant* to variations in the hydration history than the ordinary Portland cements are, so that performance specifications for concretes are becoming less concerned with strength and more with durability and long-term behaviour.

If a customer is prepared to pay more for a cement that he intends to use for a special purpose, such as for use in oil wells or for stabilizing rock-fill in mining operations, the selection and blending of raw materials and the control of the manufacturing processes can provide a wide range of closely specified cements.

Further opportunities to manipulate the properties of cement-based materials are provided by admixtures that can be added to a concrete mix. These are frequently organic materials or fillers, such as pulverized fuel ash, ground granulated blast furnace slag or finely divided powders. Typically, these may affect the workability or mouldability of the mix, the early age strength development and the durability of the product. Glass, steel or polymer fibres can also be added to concrete and mortar mixes and these also have a major effect on the properties of the material both in the fresh and hardened states. However, the hydraulic cement is the basic ingredient in the concrete formulations and in this paper a few of the controlling parameters have been discussed in an attempt to dispel the often quoted view that modern cements differ solely because of changes in either their fineness or in the relative amount of alite and belite that they contain.

There are many factors that can be controlled to produce a cement to meet specific needs. It is the specification of these needs and the means of verification that the needs are satisfied in practice that are often missing. The cement makers and the concrete producers have the expertise to make cement-based materials with widely different mechanical properties, but the specifiers are not always able to define their needs; it is hoped that this will come.

Subsequent papers elaborate on the possibilities. It is from the optimization of cement chemistry, the use of suitable admixtures or fibres in the mix formulations and from the introduction of novel fabrication techniques that the new generations of cement-based products will come. The opportunities for inorganic materials seem unlimited.

References

Bernard, J. P. 1978 Thesis, University of Paris, VI.
Bogue, R. M. 1929 *Ind. Engng. Chem. analyt. Edn* **1**, (4), 192.
Chatterjee, T. K. & Ghosh, S. N. 1980 *Wld Cem. Technol.* **11**, pp. 252–257.
Chatterji, S. 1969 *Chemy Ind.*, September, pp. 1320–1324.
Corish, A. T. & Jackson, P. J. 1982 *Concrete* **16**, 16–18.
Dalziel, J. A. D. 1980 In *7th Int. Congr. Chem. Cement*, Paris: *Paris*, no. IV, pp. 93–97. Éditions Septima.
Francis, A. J. 1977 *The cement industry 1796–1914: A history*, Newton Abbot: David and Charles.
Jeffrey, J. W. 1952 *Acta crystallogr.* **5**, 26–35.
Lawrence, C. D. 1969 *Res. Rep. Cem. Concr. Ass.* no. 19.

Le Chatelier, H. 1882*a* *C.r. hebd. Séanc., Acad. Sci., Paris* **94**, 13.
Le Chatelier, H. 1882*b* *J. Soc. chem. Ind., Lond.* **1**, 151.
Maki, I. 1974 *Cem. Concr. Res.* **4**, 87–97.
Maki, I. 1979 *Cemento*, **76** (4), 167–176.
Midgley, C. M. 1952 *Acta crystallogr.* **5**, 307–312.
Midgley, H. G. 1957 *Mag. Concr. Res.* **9**, 17–24.
Mills, R. H. 1968 *Proc. 5th Int. Symp. Chem. Cement, Tokyo*, vol. 3, pp. 74–85.
Pomeroy, C. D. 1978 *Proc. Instn mech. Engrs* **192**, 135–144.
Regourd, M. 1967 *Rev. Mat. Constr.* **620**, 167–176.
Regourd, M., Hornain, H. & Mortureux, B. 1978 *Cim. Betons Platres Chaux*, **712**, 137–143.
Richartz, W. 1968 *Proc. 5th Int. Symp. Chem. Cement, Tokyo*, vol. 3, 119–128.
Roy, D. M. & Gouda, G. R. 1973 *Cem. Concr. Res.* **4**, 807–820.
Scislewski, Z. 1975 *Cah. Cent. scient. tech. Bâtim.* no. 165, p. 1351.
Tavasci, B. 1934 *Chim. ind. appl.* **16**, 538.
Tornebohm, A. E. 1897 *Tonindustriezeitung* **21**, 1148.
Tornebohm, A. E. 1903 *Zement, Charlottenb.* **4**, 287.
Tornebohm, A. E. 1911 *Bau Mat. Kunde* **6**, 142.
Yamaguchi, C. & Uchikawa, H. 1961 *Zem.-Kalk-Gips* **11**, 497–504.

Discussion

D. D. DOUBLE (*Department of Metallurgy and Science of Materials, University of Oxford, U.K.*). Dr Pomeroy mentioned that there has been a progressive increase in alite (C_3S) contents in Portland cements over the years, presumably due to more efficient heat treatment methods during clinkering. Is the cement industry looking at the possibility of altering the phase composition of Portland cement? For example, it has been suggested that there may be energy-saving advantages in producing lower temperature belite (C_2S) rich cements.

C. D. POMEROY. The Cement and Concrete Association does not undertake research into the cement manufacturing processes, so that I am unable to answer this question. However, Dr Moir is presenting a paper at this symposium, which describes some of the changes in cement composition that are under investigation.

F. MASSAZZA (*Italcementi S.p.A. – Via Camozzi* 124, 24100 *Bergamo, Italy*). The compressive strength to tensile strength ratio of pozzolanic cements is different from that of Portland cements. Is there any reason for this behaviour?

C. D. POMEROY. I have not studied this problem, but if the use of finely divided pozzolan enables closer packing of the cementitious grains to be realized, it is possible that the elimination of larger interparticle voids could result in a higher tensile strength, whereas the total porosity governs the compressive strength. This subject is discussed by Birchall *et al.* in their paper in this symposium.

P. L. PRATT (*Imperial College, London, England, U.K.*). q.X.r.d. analysis suggests that there are 100 % (by mass) crystalline components in O.P.C. What was the glassy phase content in this cement and how would it be detected?

C. D. POMEROY. The quantitative analysis referred to by Professor Pratt appears as figure 2 in our paper, where because of the limitations imposed on the reproduction of the figure it does not include the diffraction profiles of the components used to obtain the analysis. In this particular cement the diffraction profiles of the ferrite phase were not that of a well crystallized material and our analysis is given in terms of a ferrite, the diffraction profiles of which have been broadened accordingly. Satou *et al.* (1970) have suggested that the glassy to crystalline content of the ferrite phase, extracted from five Japanese cements, is about 3:7. Our limited experience with U.K. cement suggests a ratio of 1:4. On this basis the 'glassy' ferrite content of the O.P.C. would be just over 1 %.

Reference

Satou, S., Tamura, T., Takahashi, Y. & Muto, H. 1970 Studies on the ferrite phase separated chemically from industrial cement clinkers. *Rev. 24th tech. Session, May* 1970, The Cement Association of Japan, pp. 8–13.

H. A. W. CORNELISSEN (*Delft University of Technology, Holland*). How can the tensile strength of Portland cement be increased (i.e. by the chemical composition)?

C. D. POMEROY. I have no answer to this question, though my reply to Professor Massazza may apply.

Phil. Trans. R. Soc. Lond. A **310**, 17–30 (1983) [17]
Printed in Great Britain

Cement in the 1990s: challenges and opportunities

BY G. FROHNSDORFF[1] AND J. SKALNY[2]
[1] *Centre for Building Technology, National Bureau of Standards, Washington, D.C.* 20234, *U.S.A.*
[2] *Martin Marietta Laboratories,* 1450 *South Rolling Road, Baltimore, Maryland* 21227, *U.S.A.*

[Plates 1 and 2]

Despite large gaps in knowledge of cement science, cement and concrete are the preferred materials for much civil engineering construction. As the gaps are filled, cement and concrete should become even more valuable construction materials. The gaps stem in large part from the inability to characterize cements and their hydration products in unambiguous terms.

For many reasons, there have been significant barriers to cement research. The barriers have resulted in fragmentation of research efforts among groups with less than adequate mixtures of skills. Nevertheless, the authors believe there will be a revolution in cement technology based on an integration of research efforts through cooperation on national and international levels. It will be assisted by advances in telecommunications and the application of computers to materials characterization, management of distributed data bases, the development of mathematical models representing the state of the art of cementing reactions, and education of technologists working with cement and concrete. The revolution will require the development of new standard specifications for cements, particularly standards free from unnecessary prescriptive requirements.

The challenge and the opportunity for the 1990s is to make cements and concretes more uniform, more predictable materials than they are at present.

INTRODUCTION

Society depends upon cement and concrete for the constructions that represent its largest capital investments. In 1981 the value of buildings and other constructed facilities in the United States was estimated to be $\$5.3 \times 10^{12}$; this represents about 70% of the national wealth (*Statistical Abstracts of the United States* 1981).

Perhaps because of the commodity nature of cement and concrete and their 'low technology' image, investments in cement and concrete research do not appear to be proportional to the importance of these products to society, and the effectiveness of the research activities is lessened by fragmentation. In spite of the weak support for such research, we believe that a revolution in cement technology is coming. In this paper we speculate about the changes in cement technology we expect between now and the end of the century. The revolution we foresee will result from the combined influence of computers and new methods of materials characterization on understanding the factors affecting cement performance.

Because cement and concrete are less well understood than metals such as steel and aluminium, and because they can be more easily tailored to meet special needs than wood, their potential for growth appears high. Since, for the most part, cement and concrete are likely to remain as commodities, the greatest technical need is to be able to predict their performance with a high degree of confidence. Research must, therefore, contribute to uniformity of performance and performance prediction, as well as development of new and improved materials and applications. The required research will necessarily address both cement manufacture and use.

Many of the comments made in this paper apply to concrete as well as cement, even though concrete may not always be mentioned explicitly. We shall discuss needs for scientific and technical research, improved standards, and specialized education for technologists working with cement and concrete. All of these pose challenges and opportunities.

The N.M.A.B. report on the status of cement and concrete research and development in the United States

The 1980 report on the *Status of cement and concrete research and development in the United States* (Roy 1980) began with a disturbing, but undoubtedly true statement, about R. and D. in the cement and concrete industry, 'The health of the industry 20 years from now will reflect strongly the quality and scope of the effort exerted today, but there is little evidence that their relationship is widely recognized'. (The role of new ideas in initiating and sustaining progress is shown schematically in figure 1.) The report concluded that, though research and development in cement and concrete can yield great rewards, spending for basic research on cement and concrete is minimal and it is very limited for all other types of research. Further, collaboration and flow of scientific and technological information among cement producers and users and the related industrial, governmental and academic establishments is inadequate to advance the state of the art. The unsatisfactory levels of research and development in cement and concrete reflect the fragmentation of the industry, the Federal tax structure, the generally low return on investment, and the absence of centres for relevant scientific and technological education.

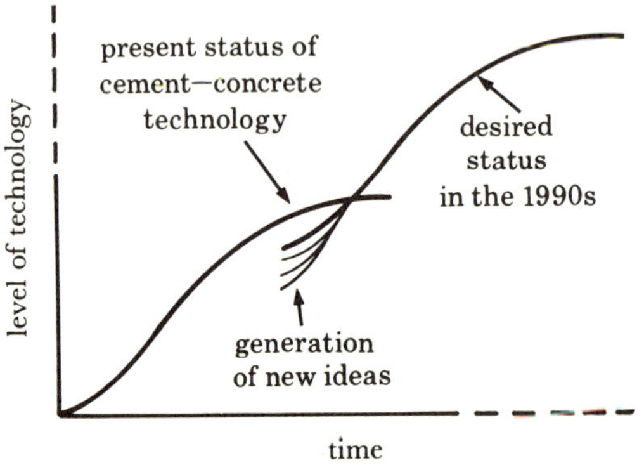

FIGURE 1. Influence of knowledge transfer on practice. New ideas, some transferred from other fields of materials science, must be reduced to practice to produce a new level of cement and concrete technology.

The report recommended that government agencies with responsibilities for energy, materials, the environment and construction should increase their support of long-range fundamental research on the manufacture and use of cement and concrete, and that special attention should be devoted to studies of (*a*) basic mechanisms, such as hydration and crystal-phase development in cement, and hardening and strength development in concrete; (*b*) long-term behaviour and durability in extreme environments; (*c*) use of energy and resources in producing cement and of utilizing concrete products; (*d*) the interaction of experiment, theory, and modelling.

A thought-provoking table indicating areas for needed improvement of concrete, table 1, was included in the report. The table draws attention to the range of uses to which cement and concrete are put; the range of properties that may have to be considered in selecting cement and concrete for a specific application; and the fact that, in various applications, cement and concrete compete with clay brick, adobe, steel, plastic, timber, aluminium and asphalt.

The challenge to the cement industry in the 1990s is to produce materials that, when used in concrete or other products, have high cost-effectiveness as compared to competitive materials. That this can be done seems evident since cement and concrete are already the materials chosen for much civil engineering construction, despite the many gaps in knowledge about factors affecting their performance. As the gaps are closed, the performance of cement and concrete in traditional and potential new uses should improve substantially.

To help the reader understand our views of the opportunities and challenges for cement in the 1990s, we shall speculate about cement (and concrete) technology as it may be 10 or 15 years from now.

Speculations about cement technology in the 1990s

Considering the changes taking place as a result of developments in computing and techniques for materials characterization, we believe the cement technology of the mid-1990s will have the following features.

(*a*) The generally accepted knowledge of the performance of cements and concretes will be stored in computer data bases and in computer-based mathematical models.

(*b*) Computer-based expert systems (Stefik *et al.* 1982) analogous to those used in medical diagnostics (Shortliffe 1976) will be available to aid designers and specifiers of concrete and other materials, thereby making the latest and best information quickly and easily available to all.

(*c*) The transfer and adaptation of basic knowledge from other fields of materials science, such as metallurgy and high temperature formed ceramics, will lead to better exploitation of the hydraulic properties of cement. (See figure 2, plate 1.)

(*d*) A branch of ceramic science and engineering dealing with formation of products by hydrothermal reactions will be developed. It will encompass cement use and its progress will be closely linked to progress in other areas involving reaction formed ceramics such as silicon nitride (Jennings 1982). (See figure 3, plate 1.)

(*e*) Improved understanding of factors affecting cement performance and more precise and meaningful characterizations of cements will greatly enhance knowledge of the relations between manufacturing conditions and cement performance. This will lead to new opportunities to optimize cement manufacturing processes in respect to energy use, product performance, and economy.

(*f*) The range of attainable mechanical properties of concrete will be much greater than is generally recognized as being possible at present. Much greater precision in predicting the performance of concrete will result in greater ability to design concrete to match the requirements for a given application.

(*g*) Greater emphasis will be placed on specifying the reliability of concretes intended to have specific service lives in specific service environments.

Table 1. Areas that need improvement in specific properties of concrete

factory fabricated units	compressive strength	flexural or tensile strength	bond strength	volume stability	controlled expansion	uniform appearance	colour	low density	flow properties	Young modulus	impact resistance	ductility	engery absorption	fracture toughness
1. block	×	.	.	×	.	×	×	×	.	.	×	.	.	.
2. brick	×	.	.	×	.	×	.	.	.	.	×	.	.	.
3. pipe	×	×	.	×	.	.	.	.	.	.	×	.	.	×
4. panels	.	×	.	×	.	×	×	×	.	.	×	.	.	×
5. beams	×	×	.	×	.	.	.	.	.	×	.	×	.	×
6. tile	.	×	.	.	.	×	.	×	×	.	×	.	.	.
7. extruded products	.	×	.	×	.	×	.	.	×	.	×	.	×	×
8. fibre-reinforced products	.	×	.	×	.	×	.	.	×	.	×	×	×	×
9. boats	.	×	.	×	.	.	.	×	.	.	×	.	.	.
10. railroad ties	.	×	.	.	.	.	.	.	.	.	×	×	×	×
field use														
1. foundations	×	×	.	.	.	.	.	.	.	.	.	.	.	.
2. missile silos	×	×	.	×	.	.	.	.	.	.	×	.	×	×
3. columns	×	×	.	.	.	×	.	.	.	×	.	.	.	.
4. slabs	.	×	.	×	.	.	.	×	.	.	.	.	.	.
5. highways	.	×	.	×	.	.	.	.	×	.	.	.	.	.
6. canal linings	.	×	.	×	.	.	.	.	×	.	.	.	.	×
7. tunnel linings	×	×	.	×	.	.	.	.	×	.	×	×	.	.
8. bridge decks	.	×	.	×	.	.	.	.	.	.	.	.	.	.
9. desalination plants	.	.	.	×	.	.	.	.	.	.	.	.	.	.
10. dams	×	.	.	×	.	.	.	.	.	.	.	.	.	.
11. marine construction	×	×	.	×	.	.	.	.	.	.	.	.	.	.
12. nuclear press vessels	×	×	.	×	.	.	.	.	×	.	.	.	.	.
13. terazzo	.	×	×	×	.	×	.	.	.	.	.	.	.	.
14. stucco	.	×	×	×	.	×	.	.	×	.	×	.	.	.
15. masonry mortar	×	.	×	×	.	.	.	.	×	.	.	.	.	.
16. oil well grouts	.	.	×	×	.	.	.	.	×	.	.	.	.	.
17. concrete patching	.	.	×	×	.	.	.	.	.	.	.	.	.	.
18. refractory linings	.	.	×	×	.	.	.	×	×	.	.	.	.	×
19. roofing	.	×	×	×	.	×	.	×	×	.	×	.	.	.
20. elevated railroad structures	×	×	×	.	.	×	.	.	.	.	×	×	.	×
21. hardened MX missile sites	×	×	×	.	.	.	.	.	.	.	×	×	×	×

early strength	quick setting	low heat liberation	low permeability	freeze–thaw resistance	sulphate and salt resistance	low thermal expansion	abrasion resistance	stain resistance	low thermal conductivity	high temperature resistance	low cost	estimate of fraction of total quantity of cement used		alternative materials
×	.	.	×	.	.	.	.	.	×	.	.	4	1.	clay brick, adobe
.	.	.	.	.	.	.	.	.	×	.	.	0.2	2.	clay brick
×	×	.	×	.	×	.	×	.	.	.	.	2	3.	steel, asbestos with cement, plastic, Techite, clay
×	.	.	.	.	.	×	.	×	.	.	.	2	4.	clay brick
×	×	.	.	.	.	×	.	.	.	.	.	2	5.	steel, lumber
×	×	.	×	×	.	.	.	×	.	.	.	0.5	6.	clay, tiles, slate, asbestos with cement, wood shingles
×	×	.	×	.	.	.	.	.	.	.	.	0.2	7.	aluminum, steel, wood, plastic
×	×	.	×	.	×	×	.	.	.	.	.	2	8.	aluminum, steel, wood, plastic, glass, fired clay
.	.	.	×	×	×	×	.	.	.	.	.	0.1	9.	steel, wood
×	×	.	.	×	×	.	×	.	.	.	×	0.5	10.	wood
.	.	×	.	.	×	.	.	.	.	.	×	40	1.	
.	.	×	×	×	×	.	.	.	.	×	.	0.2	2.	
.	.	.	.	.	.	.	.	.	.	.	.	8	3.	steel
.	.	.	×	×	×	.	.	.	.	.	.	15	4.	
.	.	.	×	×	×	×	×	.	.	.	×	15	5.	asphalt
.	.	.	×	×	×	×	×	.	.	.	×	2.5	6.	
×	×	.	×	.	×	×	×	.	.	.	×	2.0	7.	
.	.	.	.	×	×	×	×	.	×	.	.	1.5	8.	asphalt
.	.	.	×	.	×	×	×	.	×	.	.	0.1	9.	metal, plastic, glass
.	.	×	.	×	×	×	.	.	.	.	×	0.7	10.	earth fill
.	.	.	×	×	×	×	×	.	.	.	.	1.1	11.	steel
.	.	×	.	.	.	×	.	.	.	×	.	0.1	12.	steel
×	×	.	×	.	.	×	×	×	.	.	.	0.1	13.	resins
.	.	.	×	.	.	.	.	×	.	.	.	0.5	14.	
.	.	.	.	×	.	×	.	.	.	.	.	4.5	15.	resin formulations
.	.	.	.	.	×	.	.	.	.	×	.	1.4	16.	
×	×	.	.	×	×	×	×	.	.	.	.	0.1	17.	resin formulations, asphalt
×	×	×	.	.	.	×	×	.	×	×	.	0.1	18.	high alumina cement
.	.	.	.	×	.	×	.	.	.	.	.	0.1	19.	asphalt, metal
.	.	.	.	×	×	.	.	.	.	.	.	0.5	20.	steel
.	.	.	×	×	.	.	.	.	.	.	×	large	21.	steel

(*h*) Cement standards will emphasize evaluation of performance and uniformity, and prediction of performance under varying conditions. Performance specifications will revive the incentive to develop proprietary cements and concretes.

(*i*) Understanding of factors affecting the performance of cements and concretes will make possible more rational decisions about the use of waste and by-product materials such as fly ash, blast-furnace slag, and silica fume in cement and concrete. Much more of these and other similar materials will be used in concrete.

(*j*) Computer-based aids to decision-making and sophisticated quality assurance programs will lead to improved cement formulations and improved cement manufacturing operations. Improved ability to assess quality will result in stronger incentives to improve the quality and uniformity of cements and concretes.

(*k*) Advances in cement technology will require more sophisticated technologists. Large data bases and mathematical models will aid the development of special academic training programs and provide ever-present opportunities for computer-aided individual instruction.

(*l*) International technical societies, aided by advances in telecommunications and distributed computing, will play a vital role in fostering worldwide collaboration among scientists involved with cement and concrete.

If, as we believe, these possibilities are desirable, we must ask how we may expedite their attainment. Let us now look at the scientific and technical needs, and then consider the needs for standards and education.

Scientific and technical needs

The challenges and opportunities facing the large element of the construction industry that uses cement are enormous considering the gaps in knowledge of factors affecting performance of cement in concrete (Sereda & Ramachandran 1975*a*, *b*). Our comments on gaps to be filled will apply specifically to Portland cements, but analogous statements can be made about other cementitious materials.

Before discussing the gaps in knowledge related to cement use and cement manufacture, we must note the common gap of how to characterize cement, its precursors and hydration products, in unambiguous terms related to performance. Inadequate ability to characterize the materials causes much of the effort to understand the manufacture and use of cements to be wasted.

The gaps in knowledge relating to cement characterization prevent us from being able to answer such questions as: what information is needed to describe a cement adequately for the purposes of (*a*) research ,(*b*) specifications and (*c*) making decisions concerning suitability for a given application? How can the information be obtained at a reasonable cost? The magnitude of the knowledge gap is apparent from the collective inability of scientists to answer satisfactorily some obvious questions about any cement, or even to agree on the relative importance of the questions. Examples of the questions that might be asked about a cement are: what is its particle size distribution? What is its elemental composition? What phases are present and how much is there of each? What are the compositions of the phases? What are the types and concentrations of the imperfections in the various phases? How are the phases distributed between the particles of different sizes? What are the natures of the particle surfaces?

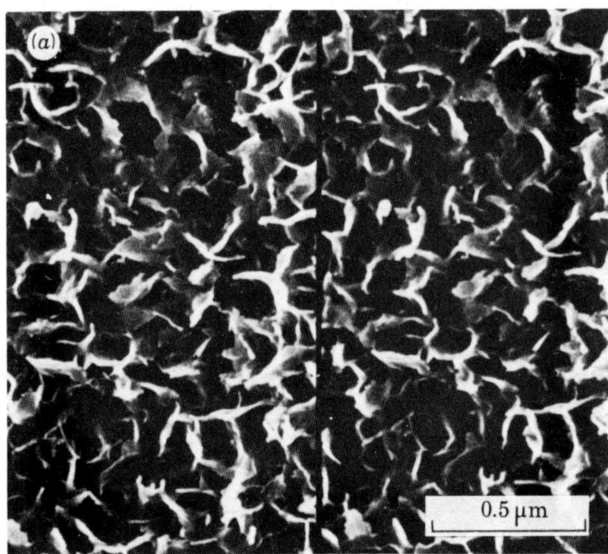

FIGURE 2. Analogies between the reactions of cement and other materials may contribute to an understanding of the reaction mechanisms and structure formation. For example, this high resolution stereo s.e.m. micrograph (*a*) and schematic drawing (*b*) of $Al(OH)_3$ show a 'honeycomb' morphology (Venables *et al.* 1980) resembling that of gel formed in the early minutes of $Ca_3Al_2O_6$ hydration (Breval 1976). Similar to hydration of cement components, Al_2O_3 hydration exhibits an induction period that can be modified by dissolved additives.

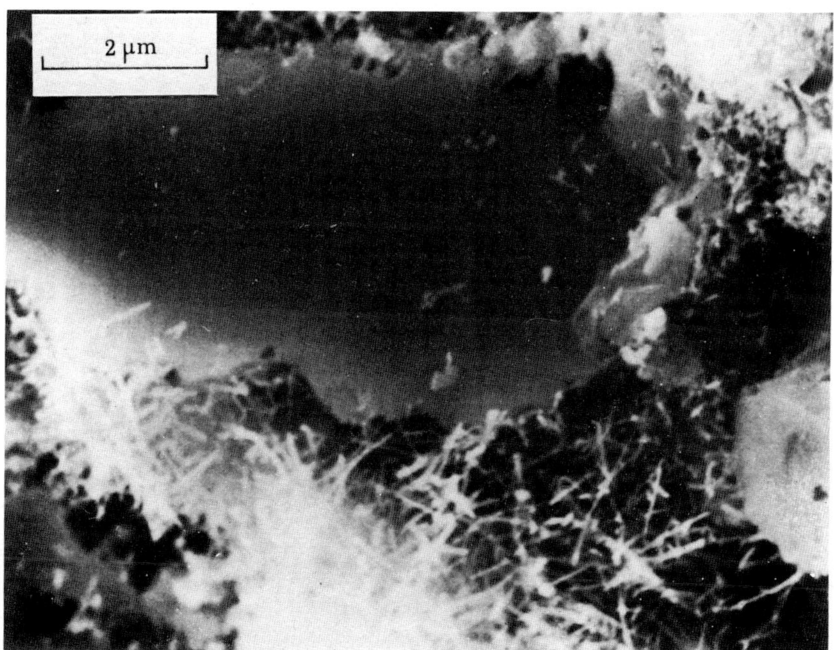

FIGURE 3. S.e.m. micrograph of silicon nitride showing a similar micromorphology of that of a cement paste at an early stage of hydration (Jennings 1982). This resemblance – as well as other similarities to reaction-formed ceramics (for example, induction periods, 'inner' and 'outer' products, brittleness) – may be the result of similar reaction mechanisms.

(*Facing p.* 22)

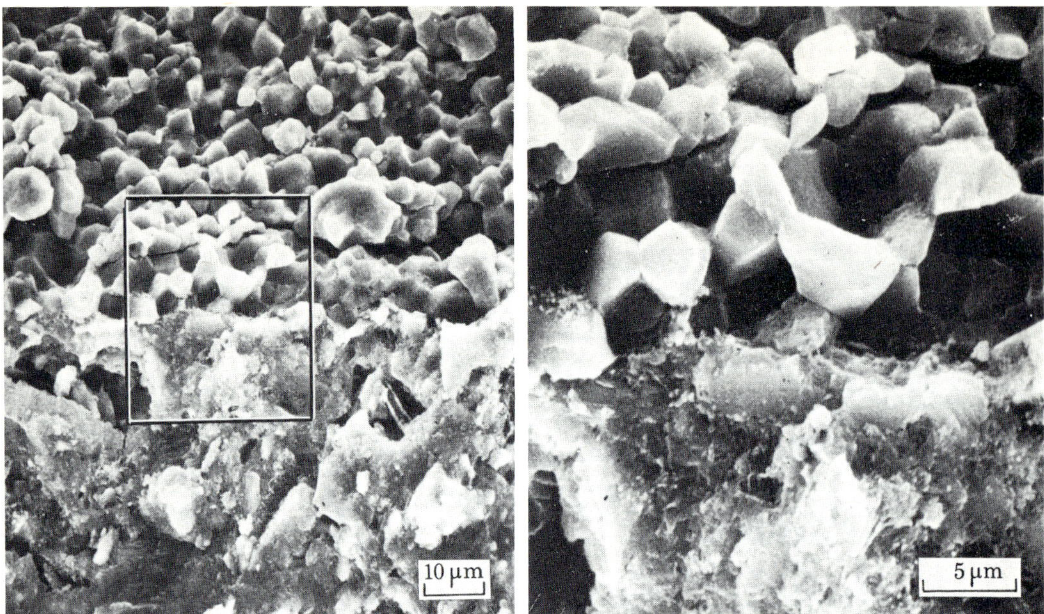

FIGURE 4. There are many unanswered questions about the cement paste–aggregate bond and its effects on concrete performance. These s.e.m. micrographs show the interface of a cement paste – quartz composite fractured normal to the interface. In contrast to samples made with limestone, calcium hydroxide cannot be seen in the 'aureole de transition' and the paste at the interface appears to have about the same porosity as the bulk cement paste (Mindess & Struble 1982).

FIGURE 6. Improvements in cements and concretes can lead to improvements in civil engineering. The Water Tower Place in Chicago is the world's tallest reinforced concrete frame building (*ca.* 260 m). The structural columns of its lower floors were constructed with 62 MPa (90 days) ready-mixed concrete, which was not available until about 10 years ago.

Cement use

There are uncertainties not only about the quantitative relations between measurable characteristics of anhydrous cements and their contributions to the engineering performance of concrete, but also often about the qualitative relations. The uncertainties particularly concern factors affecting durability and other aspects of long-term performance. There is a challenge to define the factors controlling the engineering properties of hardened cement pastes in concrete and to express the information in the form of mathematical models which can help reduce the uncertainties. This was attempted for sulphate resistance by Dunstan (1982). Examples of needed individual models, which might ultimately become part of a comprehensive macromodel of cement manufacture and use (Frohnsdorff & Clifton 1981), are models relating cement characteristics to (*a*) cement hydration and microstructure formation; (*b*) environmental effects on hardened cement pastes; (*c*) rheological behaviour of fresh cement pastes, including the effects of shear on the subsequent properties of hardened pastes; (*d*) the reactions of cement pastes with aggregates and mineral admixtures, including the chemistry and physics of the paste–aggregate bond; and (*e*) the micro- and macro-structures of hardened cement pastes and their relation to the engineering performance of concrete. The models should extend to the undoubtedly very complex and challenging examples of cement pastes and concretes containing chemical and mineral admixtures.

Examples of significant gaps in knowledge relating to cement use have been provided by recent conferences on the rheology of concrete and on the cement paste–aggregate bond. The participants in the Materials Research Society Symposium on concrete rheology (Skalny 1982) agreed that lack of knowledge of rheological properties (as well as inadequate dissemination of existing knowledge) is hindering development of concrete processing equipment and, hence, hindering the progress of concrete technology. They believed that new processing techniques, based on recent research, could lead to concrete that would be easy to place even at low water contents, and have much higher strengths than are now attainable, and have greater durability. This suggests that closing gaps in this area of knowledge should make it possible to decrease the life-cycle cost: benefit ratios of concrete structures, whether through using concrete closer to its theoretical limits, or by increasing the safety factors and, hence, the service life.

For the cement paste–aggregate bond (Struble *et al.* 1979), the common belief that the 'bond', however defined, is often the strength-limiting factor in concrete has been brought into doubt by presentations at two recent international conferences on bonding (Bartos 1982; I.N.S.I. 1982). Other opinions that the bond may not be the 'weak link' rest on the following evidence: (*a*) the paste–aggregate interface does not necessarily provide an easy pathway for water flow (Mindess & Struble 1982; Wakeley & Roy 1982); (*b*) whereas damage at or near the paste–aggregate interface, such as the formation of drying cracks, reduces strength, it has not been shown that improvement of the bond will increase strength; and (*c*) since the paste in concrete is usually confined between quite closely packed, mechanically stronger, grains of sand and aggregate, the fact that a crack propagates through, or close to, the paste–aggregate interface is not conclusive evidence of weakness of the bond (Diamond *et al.* 1982). It is clear that answering many questions relating to the nature of the bond, its importance to the performance of concrete, and the effects of the cement upon it, remains a challenge. (see figure 4, plate 2.)

Research has contributed significantly to most of the technological advances in cement and concrete. Among the recent contributions that are likely to have strong influences on the cement and concrete technology in the 1990s are knowledge of water reducing and super-water reducing admixtures (Hattori *et al.* 1964), macro-defect-free cement pastes (Birchall *et al.* 1981), concretes containing well dispersed, ultrafine particles (Bache 1981) or ground blast-furnace slag (Roy & Idorn 1982), and the cement–aggregate bond. It is noteworthy that all of these confirm the insights of earlier scientists (for example, Feret 1892, 1906; Abrams 1918; Davis 1937; Powers & Brownyard 1948; Kaplan 1961; Lott & Kesler 1966; Yudenfreund *et al.* 1972). They are evidence that opportunities abound, even though we may be slow to recognize and exploit them.

Cement production

Turning to cement manufacture, we do not yet know how to optimize the steps in cement production, individually or collectively, so as to produce the highest quality, most economical, or most uniform product (Roy 1980). Specifically, because of gaps in knowledge, we do not know with certainty: (*a*) how to define the quality of a clinker; (*b*) the relations between the mineralogy of the raw materials and the economy of their transformation into clinker of a given quality; (*c*) the effects of minor components on the clinkering reactions and clinker quality; (*d*) the effects of the kiln environment on clinker quality; (*e*) how to optimize heat transfer between burning fuel and kiln feed (see figure 5); (*f*) how to predict the effects of grinding on particle size and size distribution and how to optimize comminution processes; (*g*) how to predict the effects of time and conditions of cement storage on the performance of cements; and (*h*) what measurements of raw materials and products are needed, or how they should be made to provide feedback for optimal process control.

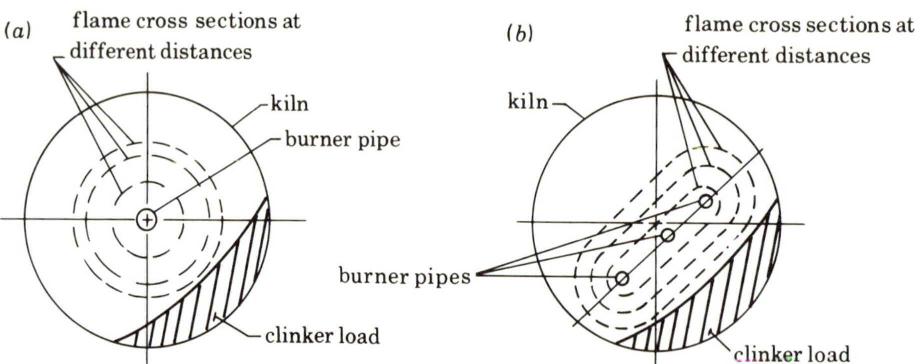

FIGURE 5. New engineering ideas could lead to improved energy efficiency in clinker production and more uniform cement. For example, the usual, approximately axisymmetric flame cross section (*a*) in rotary kilns might be replaced by an elliptical cross section (*b*) by varying the number, spacing, and orientation of the burner pipes (Weil 1983; personal communication). This should lead to better heat transfer and better control of clinker quality. The optimum geometry for such an arrangement would be arrived at by a combination of computer modelling and experimentation.

The costs of filling the gaps in knowledge of cement manufacture will be high, but the challenge must be met. We strongly believe that improvements in control of cement production processes will lead to improved cement quality, including uniformity, and better use of the chemical energy stored in cements (Idorn 1980). The benefits will include improved engineering performance of traditional products (see figure 6), better utilization of energy and natural resources, new applications of cement and, most important, reduction in life-cycle cost: benefit ratios of concrete structures.

While the filling of gaps in knowledge of cement science and technology is essential, the technology is so closely tied to standard tests and specifications (Mehta 1978) that possible modification of cement standards must now be discussed.

Cement standards

Although engineers and building officials must have basically similar performance requirements for cements, standard cement specifications for different countries are often significantly different. For example, comparison of ASTM and British standards for sulphate-resisting cements (ASTM C-150 and BS 4027) shows that it may be impossible for any single cement to meet both standards simultaneously and suggests that, in the unlikely event that one of the standards is optimal, the other must be either too restrictive or too lax. The important question, 'Which standard is better?' is difficult to resolve because, at present, there is no generally accepted performance test for sulphate-resistance nor any widely accepted basis for predicting sulphate resistance (other than compliance with these or other national standards).

For blended cements made from Portland cement clinker and either a pozzolan, such as fly ash, or a blast-furnace slag, inconsistencies among standards appear to be greater than for sulphate-resisting Portland cements. National standards differ greatly in the compositional ranges used as the basis for classification of blended cements (*Cement standards of the world* 1968).

For the most part, cement standards have both prescriptive and performance requirements. The prescriptive requirements often appear to be unnecessarily restrictive and to preclude the use of some compositions that would perform satisfactorily. For this reason, there is interest in writing certain cement standards exclusively in performance terms. While it is generally agreed that performance standards should at least be available as an option, the difficulties and the time needed to prepare them should not be underestimated. For example, whereas an editorial or other minor change in an ASTM cement standard can often be achieved within a year, many more years, perhaps 5–15, might be needed to make a really substantial change. At the risk of oversimplification, the delay occurs because it is difficult to reach a consensus among committee members when there are gaps in the data needed to make a convincing case that the benefits of a proposed change will outweight the possible risks. In the present state of knowledge, significant gaps are inevitable.

Despite the difficulties, good progress is being made towards the establishment of an ASTM performance standard for blended cements. If the activity, which has already taken eight years, is successful it should open up opportunities for manufacture of blended cements of new compositions. An essential complementary activity to the development of a performance standard is the development of needed tests, particularly tests related to durability, for example, sulphate resistance.

We believe a workable performance specification for structural cements will be established by ASTM by the 1990s, and that it will provide new incentives for cement research. It could stimulate research on new formulations, particularly those containing waste or by-product materials such as fly ash and blast-furnace slag, and also other less obvious ingredients. Further, because of the challenge of minimizing costly and time-consuming performance testing, it will stimulate improvement of performance tests and the ability to predict performance from knowledge of the cement characteristics and the expected service environment. The movement to performance specifications could aid the harmonization of cement specifications among different countries. A substantial fundamental research effort will be needed to achieve the full benefits from performance specifications, but once tangible signs of progress are seen, the potential benefits should generate increased support for research.

Education of technologists working with cement and dissemination of knowledge

Because of the complexity and multidisciplinary nature of the problems to be solved, cement research places great demands on the researcher and on those who wish to make practical use of the research results. Consistent with the low level of cement research, at least in the U.S., no institution of higher learning offers a comprehensive range of courses in cement science and technology. This situation must change if the cement and concrete industries are to achieve their full potential. Some signs of change can be seen. For example, audio-visual and computer-based teaching aids are being developed through professional societies like the American Concrete Institute and through the *Journal of Educational Modules in Materials Science and Engineering*; also there should be high educational value to the comprehensive mathematical models linking cement chemistry and physics to the physical and mechanical properties of concrete, which are being developed (Pommersheim & Clifton 1980; Popovics 1976).

Closely related to the education of technologists and the impact they will make on the cement technology of the 1990s will be their ease of access to large data bases. The progress now being made in the generation and management of data bases of all sorts (Date 1981), when coupled with ability to evaluate data against mathematical models representing the state of the art, will bring a new vitality to cement and concrete technology, partly because inter-disciplinary barriers will break down.

Improvements in education and communications are likely to be the most effective avenues for advancing the knowledge of cement science and technology. They should make it possible to coordinate the present highly fragmented, inadequate cement research efforts on an inter-national basis. An encouraging example of what can be done is provided by the recently formed RILEM Committee 68-MMH on Mathematical Modelling of Cement Hydration, which is developing conceptual models of cement hydration reactions that could become the basis for mathematical models capable of solution by computer. The potential impact on cement science appears to be great, since it is encouraging a substantial number of leading scientists, working with cement, to interact more directly and positively than ever before in the search for a unified theory of cement hydration. The knowledge gained should have a worldwide impact on cement and concrete technology by providing a basis for improving the following: techniques for predicting cement performances; tests of cement performance and uniformity; micro-structural engineering to optimize performance of cement in concrete; education of scientists and technologists; cement specifications; and research plans.

The RILEM committee cannot address all the problems that could benefit from inter-national collaboration. It can, however, indicate the potential benefits to be achieved through collaboration. These are likely to be amplified through the growing availability of computers, computer software and telecommunications.

Concluding remarks

There are many gaps in knowledge of cement science and many barriers to innovation in cement technology. Among barriers mentioned in the N.M.A.B. report (Roy 1980) were: complexity of the scientific and technical problems; lack of understanding of technical possibilities for progress; fragmentation, inadequate focus, and discontinuity of research efforts; inadequate mechanisms for productive and timely exchange of ideas among scientists and, more importantly, between basic research and users of the developed ideas; commodity nature of the cement and concrete industries and their low-technology image; improbability of short term return on investment in cement research; unsuitability of the patent system for providing incentives for investment in cement research; lack of performance-based standards and specifications; and inadequate specialized education of technologists working with cement and concrete.

Despite the gaps in knowledge and the barriers to innovation, there are increasingly good opportunities for making cement and concrete more valuable and versatile materials by improving the uniformity of cements; the ability to predict cement performance from measurable chemical and physical characteristics; and the knowledge of factors affecting cement performance, even for cements outside the current range of commercial manufacture.

Since fundamental cement and concrete research activities are so fragmented, it is questionable whether an adequate number of specialized scientists with all the requisite skills is to be found in any single organization. If this is so, some new institutional mechanisms are needed to change this. Among the possible mechanisms is increased collaboration on an international level, either informally or under the auspices of an international technical society. The collaboration would involve development of research agendas and plans, and timely sharing of research results. This idea takes on real significance at this time when specialized data bases are becoming practicable and mathematical models show promise of developing as concise representations of the state of knowledge of cementing reactions and the associated changes in physical mechanical properties. Increasing the level of collaboration poses a challenge and offers opportunities for filling the gaps in knowledge of cement science and overcoming the present barriers to progress.

We wish to thank the following persons who have contributed figures and ideas used in this paper: Dr H. Jennings, Dr S. Mindess, Mrs Leslie Struble, Dr J. Venables, and Dr J. Weil. We also wish to thank the National Academy of Sciences for permission to copy table 1 of this paper from the N.M.A.B. report.

References

Abrams, D. A. 1918 *Design of Concrete Mixtures, Bull.* no. 1. Chicago: Lewis Institute, Structural Materials Research Laboratory.

Bache, H. H. 1981 *Densified cement/ultrafine particle-based materials.* Aalborg, Denmark: Aalborg Portland.

Bartos, P. (ed.) 1982 *Bond in concrete.* London: Applied Science Publishers.

Birchall, J. D., Howard, A. J. & Kendall, K. 1981 *Nature, Lond.* **289**, 388.

Breval, E. 1976 *Cem. Concr. Res.* **6**, 129–138.

Cement standards of the world 1968 Paris: Cembureau.

Date, C. J. 1981 *An introduction to data base management systems*, 3rd edition. Menlo Park, California: Addison-Wesley.

Davis, R. E. 1937 *J. Am. Concr. Inst.* **33**, 557.

Diamond, S., Mindess, S. & Lovell, J. 1982 In *Symposium on Bonds Between Cement Pastes and other Materials, Proceedings*, pp. C. 42–C. 46. Toulouse: Institut National des Sciences Appliquées.

Dunstan, E. R. 1982 *A spec odyssey – sulphate resistant concrete for the 1980s.* ACI Publication SP–77. Detroit: American Concrete Institute.
Feret, R. 1892 *Annls Ponts Chauss.* **4**, no. 21. Mémoires Série 7.
Feret, R. 1906 *Étude expérimentale du ciment armé.* Paris: Gauthier-Villars.
Frohnsdorff, G. & Clifton, J. R. 1981 In *Cements research progress, 1980* (ed. J. F. Young), pp. 279–304. Columbus, Ohio: American Ceramic Society.
Hattori, K., Yamakawa, C. & Akitoshi, T. 1964 *On the dispersing properties of the beta-naphthalenesulfonic acid-formalin condensate for cement, Proceedings 18th General Meeting, Technical Session.* Tokyo: Cement Association of Japan.
Idorn, G. M. 1980 *The concrete future.* Holte, Denmark: Danish Concrete Institute.
I.N.S.A. 1982 *Symposium on Bonds Between Cement Pastes and Other Materials.* Toulouse: Institute National des Sciences Appliqué.
Jennings, H. M. 1982 Review on reactions between silicon and nitrogen. *J. Mater. Sci.* (In the press.)
Kaplan, M. F. 1961 *Proc. Am. Concr. Inst.* **58**, 591–611.
Lott, J. & Kesler, C. E. 1966 In *Symposium on Structure of Portland Cement Paste and Concrete*, pp. 204–218. Special report no. 90. Washington, D.C.: Highway Research Board.
Mehta, P. K. (ed.) 1978 *Cement standards – evolution and trends*, STP 663. Philadelphia: American Society for Testing and Materials.
Mindess, S. & Struble, L. 1982 Preliminary Data on the Cement–Aggregate Bond, presented at the 84th Annual Meeting of the American Ceramic Society, Cincinnati.
Pommersheim, J. M. & Clifton, J. R. 1980 In *Cements research progress 1979* (ed. J. F. Young), pp. 281–307. Columbus, Ohio: American Ceramic Society.
Popovics, S. 1976 *Cem. Concr. Res.* **6**, 343–350.
Powers, T. C. & Brownyard, T. L. 1948 *Physical properties of hardened Portland cement paste, Bulletin* no. 22. Skokie, Illinois: Portland Cement Association.
Roy, D. M. (ed.) 1980 The status of cement and concrete research and development in the United States. *N.M.A.B. Rep.* no. 361. Washington, D.C.: National Academy of Sciences, National Materials Advisory Board.
Roy, D. M. & Idorn, G. M. 1982 *J. Am. Concr. Inst.* **79**, 444–457.
Sereda, P. J. & Ramachandran, V. S. 1975 *J. Am. ceram. Soc.* **58**, 94–99, 249–253.
Shortliffe, E. H. 1976 *Computer-based medical consultations: MYCIN*, Elsevier Computer Science Library. New York: Elsevier Scientific Publishing Co.
Skalny, J. (ed.) 1983 *Materials Research Society, Annual Meeting, Symposium on the effect of Surface and Colloid Phenomena on Properties of Fresh Concrete, Boston, November 1982.* Materials Research Society,
Standard Specification for Portland Cements 1981, ASTM C–150. *Annual book of ASTM Standards,* Part 13. Philadelphia: American Society for Testing and Materials.
Standard specification for sulphate-resisting Portland cement 1982, BS 4027. London: British Standards Institute.
Statistical abstracts of the United States 1981 Washington: Bureau of Census, Department of Commerce.
Stefik, M., Aikins, J., Balzer, R., Benoit, J., Birnbaum, L., Hayes-Roth, F. & Sacerdoit, E. 1982 *Artif. Intell.* **18** (2), 135–173.
Struble, S., Skalny, J. & Mindess, S. 1979 *Cem. Concr. Res.* **10**, 277–286.
Venables, J. D., McNamara, D. K., Chen, J. M., Ditchek, B. M., Morgenthaler, T. I., Sun, T. S. & Hopping, R. L. 1980 In *12th National SAMPE Technical Conference, Proceedings*, pp. 909–923. Society for the Advancement of Material and Process Engineering.
Wakeley, L. D. & Roy, D. M. 1982 *Cem. Concr. Res.* **12**, 533–534.
Yudenfreund, M., Odler, I. & Brunauer, S. 1972 *Cem. Concr. Res.* **2**, 313–330.

Discussion

B. A. PROCTOR (*Pilkington Brothers p.l.c., Ormskirk, Lancashire, U.K.*). I have become increasingly doubtful that performance standards can, by themselves, be adequate and satisfactory. The reason for this doubt lies in the multiplicity of performance requirements required from a material and often only implied in the actual standard itself. However, in a 'recipe' standard it is known from experience that a material conforming to that description will perform a number of functions adequately in a wide variety of conditions. For example, we do not only wish to know the strength of concrete – which might be covered by a performance specification but met by a number of cements and mixes – but we also wish to know that it will retain that strength in freeze–thaw or marine conditions, not be too porous, not corrode the steel, etc. Factors covered by (say) using cement to BS 12 and a well-proven mix.

I wonder whether others share my doubts that performance standards can provide all the answers in the future, and my feeling that 'recipe specifications' still have a significant and vital role to play.

J. P. SKALNY. Mr Proctor's scepticism is well taken: we do have many technical barriers to overcome before performance specifications become generally acceptable and some aspects of specifications may (and should) remain prescriptive for a long time. However, we believe that, with increasing understanding of the relation between the composition, microstructure, and physical performance, the trend will be largely towards performance specification. The use of modern computer techniques and modelling will help in their development and acceptance.

A. KELLY, F.R.S. (*University of Surrey, U.K.*). Standards-writing bodies specifically eschew the notion of fitness-for-purpose when writing standards. The first questioner had, I thought, a very good suggestion to make, but to follow it will require a change of philosophy among standards-writing bodies.

J. P. SKALNY. I wish to comment on Dr Kelly's first point. I think this has been generally true, but that changes are taking place in the standards process. It has been customary to separate specifications from test methods in standards and to exclude fitness-for-purpose, specifically relating to durability, from the specification. For cements, this was satisfactory as long as all cements fell into a relative-narrow range of compositions with a relatively-narrow range of behaviours. The situation has to change when the need to introduce new compositions, for whatever reason, becomes sufficiently great. At this point, the comparative performance of cements of different types has to be able to be evaluated. At least in the United States, it is government policy to encourage the writing of performance specification as a way of minimizing barriers to innovation. There is no question that it is difficult, but the benefits are recognized by standards bodies, as well as governmental agencies.

F. TAMÀS (*Department of Silicate Chemistry, University of Veszprem, Hungary*). Dr Skalny mentioned the importance of bond between cement paste and aggregate. Could this somehow explain the high compressive strength of concretes with carbonaceous aggregate (crushed limestone or dolomite), as contrasted with ordinary (siliceous) aggregate, especially at relatively early ages?

J. P. SKALNY. There is still much to be learned about the bond between paste and aggregate. It is widely accepted that limestone aggregates form a reaction layer on the surfaces when in contact with Portland cement paste. We think it possible that the product of the paste–aggregate reaction could account for differences in behaviour between carbonate and silicate aggregates.

L. MCCURRICH (*Fosroc Technology Ltd, Central Laboratory, Leighton Buzzard, U.K.*). It would be interesting if Dr Skalny could speculate about developments in cement technology to improve dimensional stability. Dr Pomeroy previously spoke about cement-based machine parts but I remember that for the concrete generator stators produced some years ago, they had to be given vacuum treatment and polymer impregnation to improve long term dimensional stability.

Much of the early work on shrinkage-compensating cements originated in the U.S.A. It would be interesting if Dr Skalny could comment on their status in the U.S.A. today and on future developments in shrinkage-compensating cement.

J. P. Skalny. Dimensional changes of cementitious systems reflect changes in temperature, and changes in moisture content. This is in addition to any effects from volume changes caused by chemical reactions like those in shrinkage-compensating cements. Dealing with the drying shrinkage, it has been shown that the effects are minimized if the cement is maintained at an optimum SO_3 content since cement with an SO_3 content of 0.5 % less than the optimum could have a drying shrinkage 10–25 % greater than a corresponding cement with the optimum SO_3 content. There are no clear separations in shrinkage potential between different types of Portland cements. In one study, shrinkages for different cements ranged from about 25 % less than the median value to 40–50 % above.

With regard to the future developments in shrinkage compensating cement – I'm afraid I cannot answer the question as I am only marginally familiar with the technical aspects of the issue.

P. Jackson (*The Rugby Portland Cement p.l.c., Rugby, U.K.*). Would Dr Skalny indicate the level of expenditure (as a percentage of turnover) that the industries involved should adopt to achieve an adequate level of progress towards the objectives he has outlined?

J. P. Skalny. This is a difficult question to answer, but the level of expenditure should certainly be well above the presently estimated 0.1 % of sales (0.01 % for basic R. and D.). And it is not only the funds! More importantly, long-term research needs good management, stability, and moral support from above (corporate and governmental).

Phil. Trans. R. Soc. Lond. A **310**, 31–42 (1983) [31]
Printed in Great Britain

Cement in the context of new materials for an energy-expensive future

By J. D. Birchall, F.R.S.
Imperial Chemical Industries p.l.c., New Science Group, Runcorn, Cheshire, England, U.K. and The University of Surrey, Guildford, Surrey, England, U.K.

[Plate 1]

Hydraulic cements are energy-cheap relative to other common materials, are manufactured on a large scale and, when mixed with water, form readily mouldable pastes that harden at low temperature. In a technological sense, such pastes can be regarded as inorganic 'plastics', but the types of article that can usually be fabricated from the cements has been restricted by the low tensile strength and fracture toughness of hardened cement pastes. Poor mechanical properties are not inherent in inorganic solids formed under mild conditions; mineral structures of biological origin can display relatively high strength and useful toughness as a result of microstructural features determined by biopolymers. Recent studies have shown that the low tensile properties of cement paste result from the presence of macroscopic pores. The elimination of such defects by the use of polymeric rheology modifiers gives unreinforced cement pastes a flexural strength of 150 MPa or more. Such novel materials should considerably extend the range of uses for hydraulic cements.

Introduction

The hydraulic cements (Portland cement, its variants and the calcium aluminate cements, etc.) are of interest, not only because of their familiar use in construction, but because they may form the basis of novel materials for a future in which energy is expensive and hence materials of high energy content are costly. Table 1 sets out the common materials of fabrication in order of total energy content per unit volume relative to ordinary Portland cement. O.P.C. is a useful standard because it is made on a huge scale (10^9 t/year) and is essentially a processed chemical, despite which it remains attractively cheap in energy terms (Duckworth 1982).

The low energy content of cement-based materials relative to plastics and metals is not their only attraction. The growth in technology in the 1950s and 1960s created a considerable interest in inorganic polymers because of their potential incombustibility and inertness. Although the term 'inorganic polymer' was initially used to define a macromolecule (usually linear) not having a backbone of carbon atoms (Currel & Frazer 1969), the term has been used to cover such substances as diamond, graphite and silica (Holliday 1970) and even cement (Cherkinski 1965). The search for the non-carbon backbone polymer has not generated many industrial products (other than the silicones) and yet the target of easily manipulated inorganic materials is becoming increasingly relevant as incombustibility is demanded (Ray 1978).

The rapid growth in the use of synthetic organic polymers was due in part to the relative ease and speed by which they could be shaped into final articles. From this view point, the hydraulic cements are also attractive; when mixed with water, cements form mobile pastes that can readily be moulded or cast, with setting taking place at ordinary temperature. Unfortunately, the set material has a number of mechanical deficiences when compared to the usual materials of fabrication. Figure 1 compares the Young modulus, flexural strength and fracture

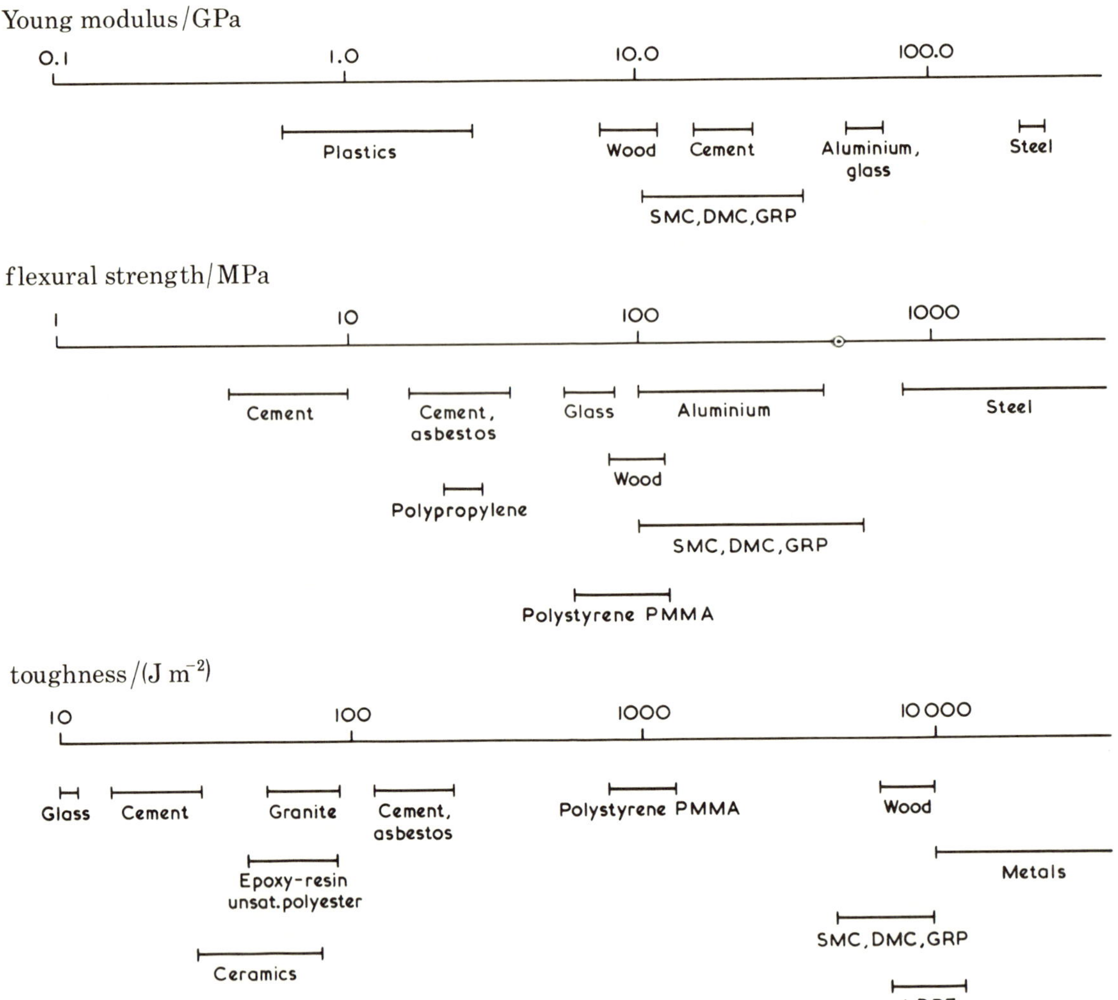

FIGURE 1. A comparison of the physical properties of common materials.

TABLE 1. TOTAL ENERGY CONTENT OF MATERIALS PER UNIT VOLUME RELATIVE TO PORTLAND CEMENT

material	total energy (volume basis)
Portland cement	1.0
flat glass	3.0
PVC	3.8
LDPE	4.2
HDPE	4.4
polystyrene	6.0
steel	19.2
stainless steel	28.8
aluminium	31.8
zinc	34.8

toughness for a variety of common materials from which the poor tensile performance and low fracture toughness of unreinforced cement products are obvious.

Thus, although hydraulic cement paste is, in a technological sense, an inorganic 'plastic' in allowing facile moulding and fabrication, the poor mechanical performance of the unreinforced hardened material has precluded its penetration into uses presently dominated by the metals and synthetic polymers. However, recent studies have shown that these deficiences are not fundamental and considerable advances have been made in the manipulation and performance of cements (Birchall *et al.* 1981*a*, 1982*a*).

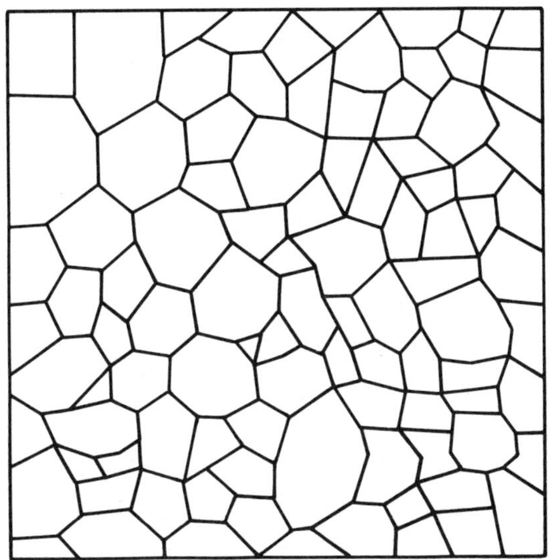

FIGURE 2. Ideal polycrystalline ceramic microstructure.

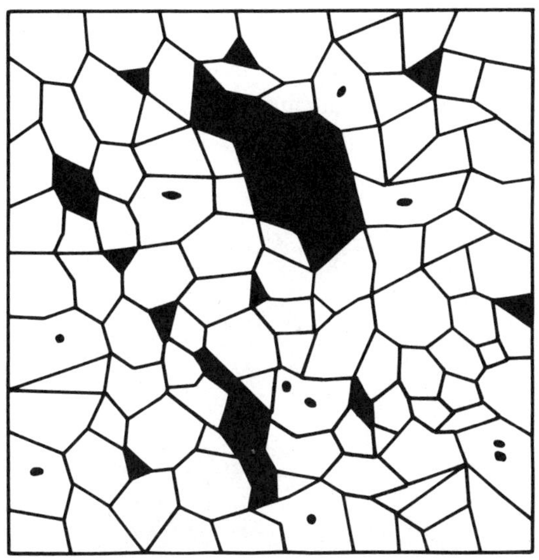

FIGURE 3. Porous polycrystalline ceramic microstructure.

APPROACHES TO THE PROBLEM

Ceramics

Once the problem has been formulated in terms, not of improving conventional cement products as such (although this may be a consequence of any new approach) but in terms of using hydraulic cements as raw materials from which *advanced* materials can be made, it becomes essential to seek new approaches in apparently unrelated areas of science. The areas that have proved to be of value are ceramics and biogenic inorganic materials. The anhydrous cements (the calcium silicates and the calcium aluminates) are themselves ceramics with bond energies of the order of 0.4 MJ mol^{-1}. Figure 2 shows an ideal conventional ceramic microstructure.

The ceramist will aim for a dense, non-porous structure in which there are no grain or phase boundary defects and in which the grain size is small; strength will rise as grain size is reduced (Petch 1953) and fall as the Griffith (1920) crack size increases. Figure 3 shows a ceramic microstructure that would give a weak material. Maximum density and minimum porosity – those features essential for strength – are attained by careful sintering at high temperature.

For the hydraulic cements, union between grains is attained by reactions with water that produce hydrated solid phases in the form of colloidal gel-like material, which itself has a high surface area (10^2–10^3 $m^2\ g^{-1}$) (Diamond 1976). An excess of water is needed for the suspension to be sufficiently mobile for casting, and this excess water is lost by evaporation. This, together with the facts that the cement particles are poorly dispersed and are extensively flocculated (and hence pack inefficiently) and that air is occluded in the suspension, results in macroscopic porosity. The *overall* porosity of the hardened material will normally be in the range 25–35 % (by volume) and will comprise pore sizes from a few nanometres (gel pores) to a few millimetres. Figure 4 is an illustration of the microstructure of a Portland cement paste.

Although there have been many attempts to relate the strength of hardened cement paste to total porosity volume, it has been shown (Birchall *et al.* 1981*b*) that the major determinant of strength is the size of the macropores resulting from poor particle packing and air occlusion. The topic is treated in greater detail by Kendall *et al.* in this Seminar. One feature that distinguishes cement paste from ceramics is the wide range of pore sizes present in cement, and confusion has resulted from the fact that although gel pores contribute vastly to overall porosity, a small number of large pores largely determine strength. In the presence of large pores, the microchemistry and morphology of the gel phase can have only a minor influence on tensile strength (Kendall *et al.* this Seminar; Birchall *et al.* 1981*b*; Alford 1981; Alford *et al.* 1982).

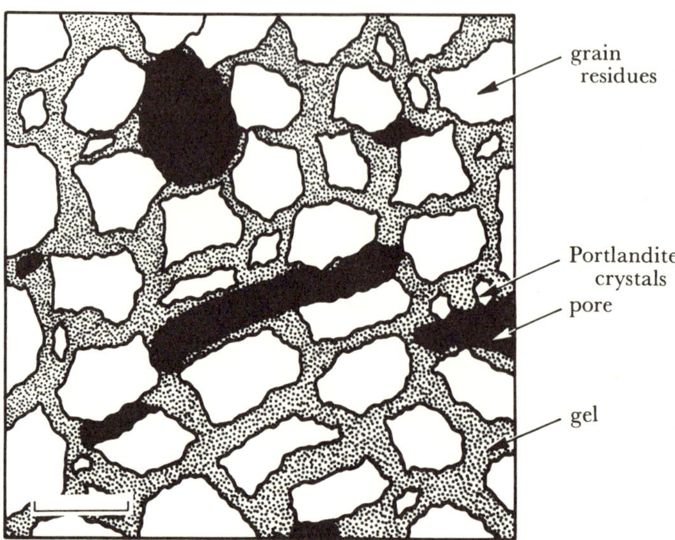

FIGURE 4. Portland cement paste microstructure. (The scale bar represents 30 μm length approximately.)

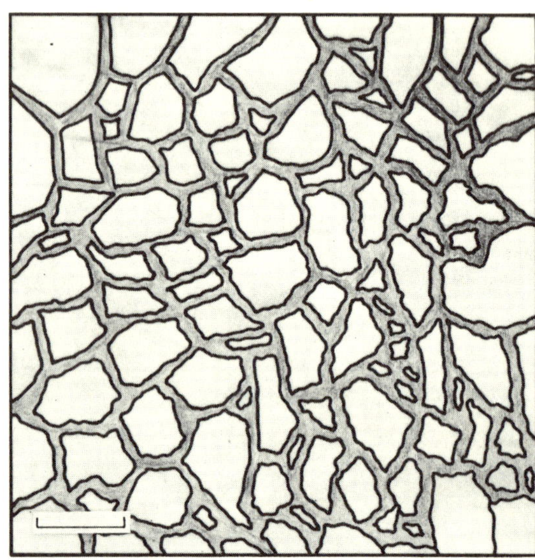

FIGURE 5. Ideal hydraulic cement paste microstructure. (The scale bar represents 20 μm length approximately.)

The ceramist would presumably argue that an ideal cement paste structure should be that shown in figure 5, with the anhydrous grains packed as closely as possible, no macroscopic defects and with a minimum of the intergrain hydrate gel within which, at least initially, are chemical bonds lower in strength by one order of magnitude than those in the grains.

It has long been known that the strength (both tensile and compressive) of cement paste rises as the ratio of water to cement (w/c ratio) is reduced (Feret 1897) but, of course, there is a practical limit to this in that a water–cement mixture becomes unworkable at low w/c ratios. The flocculated nature of cement particles does not permit fluid suspensions unless they are dilute. To an extent, this can be overcome by the addition of dispersing agents to the water, usually organic polyanions such as the products of condensing sulphonated napthalene or melamine with formaldehyde. These, by adsorption on the particles, promote dispersion and allow a workable fluidity to be retained at lower water contents (Malhotra *et al.* 1978).

However, such methods do not result in sufficient improvements in tensile strength to make cement solids suitable to replace the more conventional materials of fabrication and, indeed, 20 MPa would be a high flexural strength for a material so derived. Much macroscopic porosity remains as a result of air entrainment and defective particle packing. Figure 1 suggests that a target flexural strength of at least 100 MPa should be the aim if cement products are to compete with conventional materials.

FIGURE 6. The microstructure of mother-of-pearl (nacre).

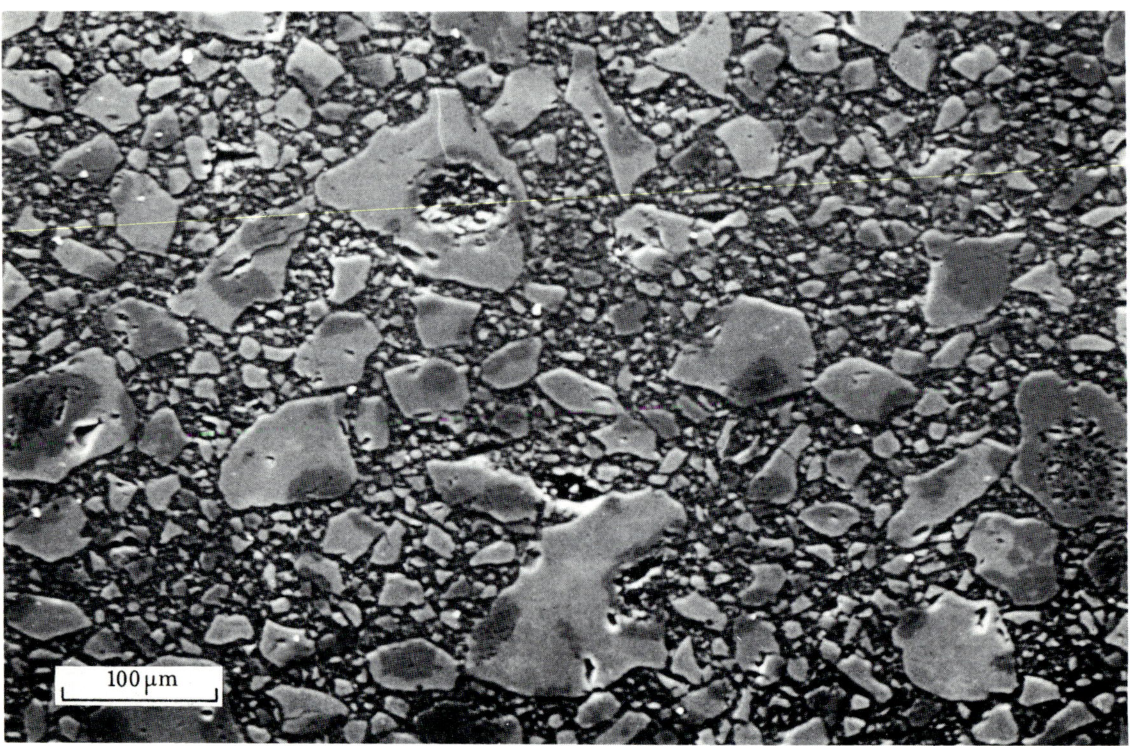

FIGURE 9. The microstructure of MDF (calcium aluminate) cement.

(*Facing p.* 35)

Biogenic inorganic materials

A study of inorganic materials of biological origin (such as skeletal structures) shows that inorganic materials formed in an aqueous system at low temperature need not be poor mechanically (Wainwright *et al.* 1976). Mother-of-pearl (nacre), which is 99% (by mass) $CaCO_3$, can have a flexural strength in excess of 100 MPa, say tenfold that of cement paste (Currey 1977). The microstructure of nacre is shown in figure 6, plate 1.

The crystallites are flat plates of aragonitic $CaCO_3$ (0.2–0.3 μm thick) arranged in layers with a tenuous interlayer of protein. Nacre has a fracture toughness comparable to that of poly-(methylmethacrylate) and this possibly results from separation of the brittle platelets as a crack advances through the matrix. The material shows marked, though not extensive, plastic deformation before fracture (maximum strain 0.018), presumably caused by plastic deformation of the organic matrix. The organic matrix in inorganic biogenic materials controls the nucleation, growth and morphology of the inorganic crystallites and ultimately, the macroscopic morphology of the structure (Nakahara & Bevelander 1971), although only a small proportion of organic matrix is necessary for this; organic matrix in excess of this can have an important mechanical role (Wainwright *et al.* 1976). The effect on nucleation is probably brought about by groups on the biopolymer which bind specific ions and promote heterogeneous nucleation at predetermined sites (Hughes 1981).

A unified approach to strong solids from hydraulic cement

If it is accepted that the ideal structure for a strong cement paste is as illustrated in figure 5, how might such a structure be made? The problem is to attain maximum particle packing, a minimum total porosity volume and the elimination of macroscopic voids. One recent approach has been to disperse the cement grains at low w/c ratio by the use of relatively high levels of a dispersing agent, such as a sulphonated naphthalene formaldehyde condensate, and to densify the intergrain hydrate gel by the addition of fine silica (average diameter 0.1 μm) (Bache 1981; Hjorth this symposium).

The compressive strength of concretes made from such materials is well over 100 MPa and, when made with strong (in compression) aggregates (e.g. calcined bauxite), can exceed 250 MPa. Tensile failure is brittle, the failure stress being about one tenth the compressive strength.

The optimum packing of particles in an aqueous phase requires more than an initial dispersion of the particles. As a suspension becomes concentrated, particles come into contact and bridge, and interparticle friction is such that an adjustment to optimum packing by sliding cannot readily occur, even under pressure (Gray 1968).

Recent work has shown that both dispersion and particle lubrication can be achieved by the use of water-soluble organic polymers – such as hydroxypropylmethylcellulose, hydrolysed poly(vinylacetate) – in a cement/water mixture (Birchall *et al.* 1981*a*, 1982*a*). A typical composition of such a mixture would be 100 parts (by mass) cement, 7 parts (by mass) polymer and 10 parts (by mass) water. The polymer–water composition alone is, of course, a stiff gel. Such a composition, on high shear mixing, forms a readily deformable dough in which there is some occluded air and in which the volume fraction of cement particles is about 0.6 (the volume fraction for dense random packing of uniform spheres is 0.63). The dough can be processed by conventional plastic techology – press-moulded, extruded, calendered, etc.

Figure 7 shows a sheet of dough that has emerged from a roller mill; the 'plastic-like' consistency of the material is obvious. Once formed into the desired shape, the object becomes hard as a result of the normal inorganic hydration reactions which, of course, may be accelerated by heat. Figure 8 illustrates the role of the polymer in the formation of a plastic cement dough.

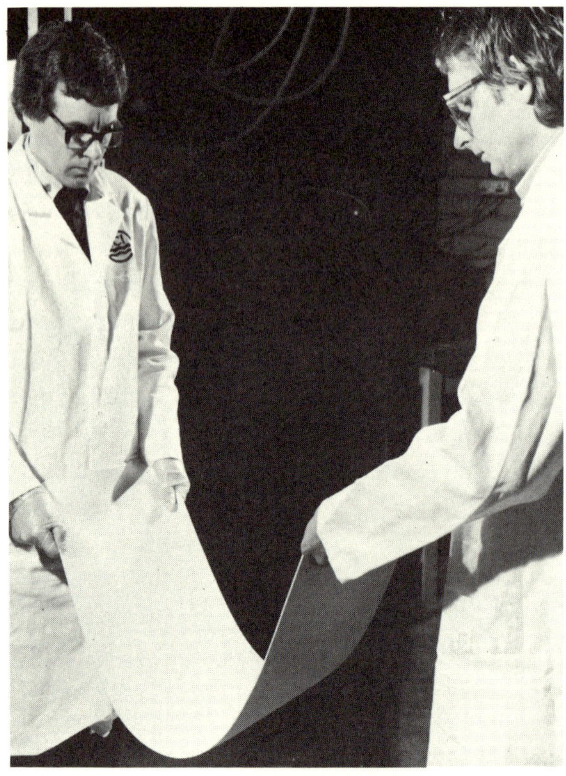

FIGURE 7. Roller-milled sheet of MDF cement dough illustrating 'plastic-like' quality.

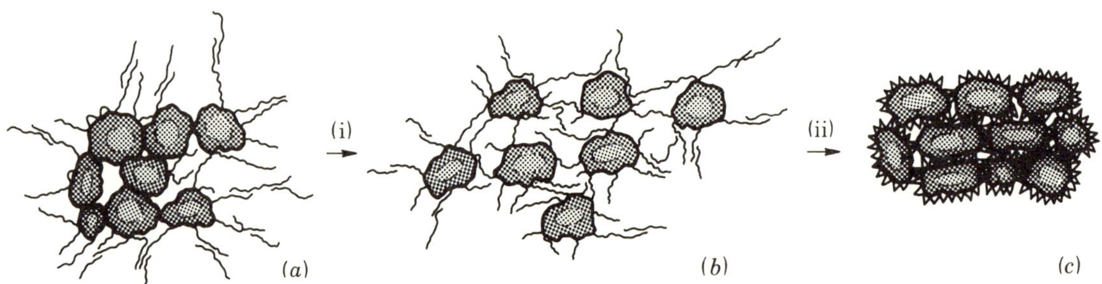

FIGURE 8. The role of the polymer in the formation of MDF cement doughs. (i) Shear; (ii) hydration of particles, dehydration of polymer.

It appears to be essential that segments of the polymer are capable of strong attachment (figure 8*a*) to the cement particles with trains of polymer extending into the aqueous phase and there entangling. It is found that intense shear mixing can be applied to such a material without cavitation and with the result that flocculated particles are torn apart (figure 8*b*). The adsorbed hydrated polymer acts as an effective interparticle lubricant, allowing sliding adjustment of the particles to close packing. The polymer remains at the interfaces between the packed particles and becomes intergrated with the inorganic hydration products (figure 8*c*).

This technique is the basis of the high flexural strength O.P.C. pastes first reported in 1981

(Birchall *et al.* 1981*b*). The aqueous chemistry of the cement and the nature of the polymer will obviously influence the rheological behaviour of the cement–polymer–water system. The polymer must, for example, be compatible with the high pH of cement aqueous phase and must not be prematurely precipitated by reaction with dissolved species derived from the cement, for example carboxylated polymers precipitating as a result of the formation of calcium salts. One specific example of a very effective system is calcium aluminate cement – poly(vinyl alcohol/acetate) – water. There is strong interaction between cement particles and polymer, and the dough formed by high shear mixing consists of packed particles (*ca.* 0.6 volume fraction) interconnected by entangled hydrated polymer chains. Gentle pressure on such a dough (less than 5 MPa) compresses occluded air bubbles and raises the particle volume fraction to about 0.7 (the volume fraction for hexagonal close packing of uniform spheres is 0.74).

If modest temperature is applied to the dough with pressure maintained, the hydration of the cement is accelerated, the deformability of the dough markedly decreases and air pockets are not reformed on release of the pressure. There is then a post-forming volume contraction (*ca.* 10 %), which raises the particle volume fraction, calculated from the original volumetric composition and measured porosity, to about 0.80. This appears to result from the contraction of the extended polymer chains that interconnect the particles as the polymer dehydrates, polymer dehydration being caused by the hydration of the inorganic cement. The inorganic hydration products increase still further the volume fraction of inorganic solid so that the final porosity volume fraction is 0.01 or less. The contraction of a polymer on dehydration can result in massive forces on solids to which the polymer is attached. The contraction of a gelatine film on glass is sufficient to tear pieces from the glass surface and this technique was used some years ago to make ornamental 'frosted' glass (Smith 1925; Gulati & Hagy 1982). The final hardened cement has a flexural strength of 150 MPa or more and the Young modulus is in the range 40–50 GPa; these properties reflect the absence of macroscopic voids and the low total porosity volume (Kendall *et al.* this Seminar). That the major role of the added polymer is as an aid to the reduction of porosity is illustrated by the fact that identical cement–polymer–water compositions can be manipulated to give both high and low strength products, only if the manipulation is such that macropores are removed is high strength achieved. Under this circumstance, it is significant that the addition of a low modulus polymer increases the modulus of the composite. The absence of macropores in high strength material is illustrated in figure 9, plate 1.

In the final densified material, the polymer resides at the interface between the close-packed cement grains and cannot be easily leached out, indicating firm binding to the hydrated material present at the interface. The initial adsorption of loops of polymer onto the particle surface may be due to 'water bridges' between the water molecules of the hydration shell surrounding cations and polar groups on the polymer (Parfitt & Greenland 1970), but it seems likely that such relatively weak bonds are replaced by strong bonds as hydration products are precipitated.

The space available for the growth of hydration products is restricted in the case of close-packed, low porosity, MDF cement and this appears to influence the morphology of hydration products. Indeed, it has been suggested that the morphology of hydration products is dependent on the space available for growth (Birchall *et al.* 1978). In macro-defect-free O.P.C. pastes, for example, there is evidence that much of the calcium hydroxide is present in microcrystalline form, the platelets being *ca.* 10 nm thick (Groves 1981). It has been suggested that large crystals of calcium hydroxide may act as flaws in the paste (Dalgleish *et al.* 1980). The absence of large crystals in the gel phase of MDF paste may contribute to the observed high tensile strength of such materials.

It will be seen that the approach to high strength cement products described above combines lessons taken from ceramic science and the science of biogenic inorganic materials – particle packing and porosity removal, and the control of microstructure by polymeric material respectively.

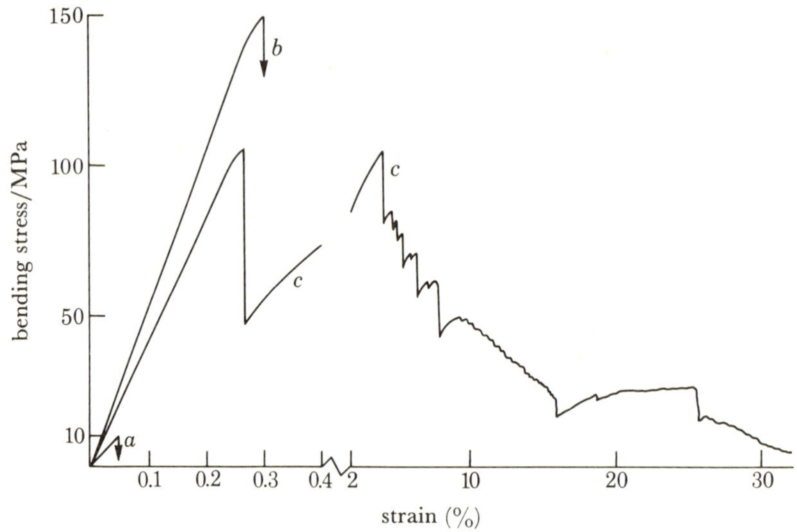

FIGURE 10. A comparison, for the three-point bend configuration, of normal cement paste, (*a*) an MDF calcium aluminate cement (*b*) and the latter reinforced with 9% (by volume) Kevlar fibre (4.5% (by volume) in the direction of maximum stress) (*c*).

APPLICATIONS

The values of the Young modulus, flexural strength and fracture toughness obtained for the best unreinforced MDF materials (50 GPa, 150 MPa and 1 kJ m^{-2} respectively) can be put into context by reference to figure 1. These properties are allied to processing and fabrication technology extensively used in the plastics industry. The levels of toughness achieved in the unreinforced material allow it to be machined easily. Fibre-reinforced materials have been prepared, for which fracture toughness approaches 10^5 J m^{-2} (N. McN. Alford, personal communication). These advances are dramatically illustrated in figure 10 in which stress–strain plots (three-point bend test configuration) are compared for conventional cement paste, an MDF calcium aluminate cement and the latter reinforced with 9% (by volume) Kevlar fibre.

Just as an organic polymer may be used alone or combined with a particulate filler, MDF paste may be filled with particulate material such as sand, silicon carbide, alumina, metal powders, etc. thereby altering secondary properties such as hardness, abrasion resistance, thermal and electrical conductivity and so on. The materials may be pigmented with organic or inorganic pigments, painted or metallized. Not surprisingly, in view of the low porosity, the rate of water penetration into MDF cements can be lower than that observed for conventional pastes by several orders of magnitude.

This technology is clearly capable of still further development as fundamental understanding increases and as experience is gained of the performance in service of these materials. Experimentally, various forms such as pipe, rod, T-section, sheet and complex shapes have been fabricated by extrusion, press moulding, calendering, and machining, and these are receiving exhaustive performance testing. There is every indication that cement-based materials will, in the not too distant future, be used in some applications at present dominated by plastics, metals and ceramics used at low temperature (Birchall *et al.* 1982*b*, *c*).

References

Alford, N. McN. 1981 *Cem. Concr. Res.* **11** (4), 605–610.
Alford, N. McN., Groves, G. W. & Double, D. D. 1982 *Cem. Concr. Res.* **12** (3), 349–358.
Bache, H. H. 1981 Dense cement ultrafine particle-based material. In *Proc. 2nd Int. Symp. on Superplasticizers in Concrete, Ottawa, Canada*. Ontario, Canada: CANMET, Dept. of Energy, Mines and Resources.
Birchall, J. D., Howard, A. J. & Bailey, J. E. 1978 *Proc. R. Soc. Lond.* A **360**, 445–453.
Birchall, J. D., Howard, A. J. & Kendall, K. 1981*a* *European Patent* no. 0021682 (priority date 26 June 1979).
Birchall, J. D., Howard, A. J. & Kendall, K. 1981*b* *Nature, Lond.* **289**, 388–389.
Birchall, J. D., Howard, A. J. & Kendall, K. 1982*a* *European Patent* no. 0055035 (priority date 22 December 1980).
Birchall, J. D., Howard, A. J. & Kendall, K. 1982*b* *Chemy Br.* **18** (12), 860–863.
Birchall, J. D., Howard, A. J. & Kendall, K. 1982*c* *Proc. Br. Ceram. Soc.* **32**, 25–32.
Cherkinski, Y. S. 1965 *Russ. chem. Rev.* **34**, 451.
Currel, B. R. & Frazer, M. J. 1969 *R. Inst. Chem. Rev.* **2**, 13.
Currey, J. D. 1977 *Proc. R. Soc. Lond.* B **196**, 443–463.
Dalgleish, B. J., Pratt, P. L. & Moss, R. I. 1980 *Cem. Concr. Res.* **10**, 665–676
Davidge, R. W. 1979 *Mechanical behaviour of ceramics*. Cambridge University Press.
Diamond, S. 1976 In *Proc. Conf. on Hydraulic Cement Pastes: their Structure and Properties*, University of Sheffield, 8–9 April 1976, pp. 2–23. Cement and Concrete Association.
Duckworth, E. 1982 *Chemy Ind., Lond.* **21**, 847–854.
Feret, R. 1897 *Bull. Soc. Encour. Ind. natn., Paris* **2**, 1604.
Gray, W. A. 1968 *The Packing of Solid Particles*. London: Chapman and Hall.
Griffith, A. A. 1920 *Phil. Trans. R. Soc. Lond.* A **221**, 163.
Groves, G. W. 1981 *Chem. Concr. Res.* **11** (5/6), 713–718.
Gulati, S. T. & Hagy, H. E. 1982 *J. Am. ceram. Soc.* **65**, (1), 1–5.
Holliday, L. 1970 *Inorg. Macromolec. Rev.* **1**, 3–16.
Hughes, M. N. 1982 In *Inorg. Biochem.* **3**, 75, Roy. Soc. Chem.
Malhotta, V. M., Berry, E. E. & Wheat, T. A. (ed.) 1978 *Proc. Int. Symp. on Superplasticizers in Concrete*, vol. I. Ontario, Canada: CANMET, Dept. of Energy, Mines and Resources.
Nakahara, H. & Bevelander, G. 1971 *Calc. Tiss. Res.* **7**, 31–45.
Parfitt, R. L. & Greenland, D. J. 1970 *Clay Miner. Bull.* **8**, 305–315.
Petch, N. J. 1953 *J. Iron Steel Inst.* **174**, 25–28.
Ray, N. H. 1978 *Inorganic Polymers*. New York: Academic Press.
Smith, W. 1925 *Glass Ind.* **12**, 277.
Wainwright, S. A., Biggs, W. D., Currey, J. D. & Gosline, J. M. 1976 *Mechanical design in organisms*. London: Edward Arnold.

Discussion

K. I. Majid (*Department of Civil and Structural Engineering, University College, Cardiff, U.K.*).

(1) Can Professor Birchall give more details about the polymer used?

(2) Can we buy it and mix our own cement to produce our own paste?

(3) Can MDF cement be used in the construction industry as ordinary cement is used in reinforced concrete structures? I am interested in its use in this field and would like to know the reasons why it cannot be used in the construction of structures and also whether we can work together in this field to ensure that it is introduced into the construction industry.

J. D. Birchall. Details of the preparation of MDF cements are given in European Patents no. 0021682 and no. 0055035 and, of course, the details revealed therein could be repeated in your laboratory, bearing in mind the proprietary interests.

It is too early to say what the future of MDF cement is in the construction industry. The process requires very much more quality control than I believe to be possible on site. There is, however, considerable interest in the use of MDF sheet as cladding or permanent shuttering, where the high strength and low permeability of the material may be an advantage.

B. W. Staynes (*Department of Civil Engineering, Brighton Polytechnic, U.K.*). When the MDF dough is being produced, air must be entrained unless deliberate steps are taken to eliminate it. Is it necessary to prepare MDF dough under vacuum?

J. D. Birchall. It is not essential to prepare MDF dough under vacuum. Air can be removed by compression of the dough.

A. A. Rahman (*Aberdeen University, U.K.*). Professor Birchall has discussed the low (less than 1 %) porosity of the MDF material. What is the level of permeability of this material for different gases and vapours? Is the residual porosity in the form of connected networks or has Professor Birchall any evidence of isolated extremely small pores or voids? Has he any information about the pore size distribution of the residual porosity?

J. D. Birchall. The oxygen permeability of a typical MDF cement is four orders of magnitude less than that of a conventional paste and the rate of water penetration six orders of magnitude less than ordinary paste. There is as yet little information about the nature of the residual porosity available.

N. J. Dave (*University of Salford, U.K.*). Professor Birchall has already mentioned that his MDF cement is not most suitable for site mixing, perhaps he could tell us whether it is suitable for various building and construction products (factory made) like boards, sheets, tiles, etc. or whether he envisages the use of MDF cement for some special applications. In brief, he could perhaps comment on the field of application of MDF cement. Has he done tests on creep and fatigue behaviour of MDF cement products? What will be the cost of MDF cement products?

J. D. Birchall. There is much interest World-wide in the potential of MDF cement in the form of sheet, tile, board, etc. Such forms have been produced and are undergoing the standard tests for such building products. I cannot give a brief reply on the question of costs relative to products made from O.P.C. because the final product cost is critically dependent on scale of manufacture and the nature (geometry, etc.) of the product.

An extensive program to study creep and fatigue is underway.

H. F. W. Taylor (*University of Aberdeen, Scotland, U.K.*). In MDF cement, a substantial proportion of the cement is presumably not reacting and acts as a micro-aggregate. It is a relatively expensive micro-aggregate. Would it be possible to replace the cement in part by some cheaper material, for example a waste material such as fly ash?

In a related question, to what extent is the glue between the unreacted cement particles provided by the cement hydration products, and to what extent is it provided by the water-soluble organic polymer?

J. D. Birchall. We have shown it to be possible to dilute MDF by the addition of fillers such as sand and fly ash to reduce the cost. Equally, it is possible to fill with rather more exotic materials such as alumina, silicon carbide, etc. to alter secondary properties such as hardness, abrasion resistance, etc.

The nature of the bonding between cement grains can be varied depending on the exact process condition but in general we regard the glue to be an intimate combination of organic polymer and inorganic hydration products.

F. TAMÀS (*University of Veszprem, Department of Silicate Chemistry, Hungary*). What is the role of plastics when preparing MDF cement? Is it only a lubricating agent? Could it be replaced by inorganic 'lubricants', montmorillonite, for example?

J. D. BIRCHALL. The polymer acts as a dispersant by transmitting shear to aggregates so that they are broken down. It then lubricates individual particles to allow adjustment to close packing. We have not found colloidal inorganic suspensions capable of this.

A. J. MAJUMDAR (*Building Research Station, Watford, U.K.*). Could Professor Birchall comment about the ease or difficulty of incorporating fibres in the MDF cement?

J. D. BIRCHALL. It is difficult to incorporate a significant volume fraction of short fibre into MDF dough without a loss of the rheological characteristics that permit facile moulding. However, we have developed methods for the production of layer laminates and these processes are cheap and effective.

A. CHAMPION (*EI du Pont Experimental Station, Wilmington DE* 19809, *U.S.A.*). In fibre reinforced. MDF cement, what degree of fibre to matrix bonding was achieved? Was there significant fibre pullout? Would there be a benefit in increasing fibre to matrix bonding?

J. D. BIRCHALL. The shear stress between Kevlar fibre tows and the MDF cement matrix measured in pull-out was found to be between 4 and 6 MPa. Pull-out measurements on a single filament have not been done.

We observed significant fibre pull-out. Increasing the fibre to matrix bond would probably improve the stress at first crack. However, our aim has been to improve toughness and for this purpose there may be little to be gained from increased bonding.

DELLA M. ROY (*Materials Research Laboratory, The Pennsylvania State University, Pennsylvania* 16802, *U.S.A.*). First, I should like to comment on the water sensitivity of the MDF cements. In our earlier work on compacted and warm pressed cement (completely inorganic) having similar strengths and very low porosities we found no problem with water sensitivity but, in fact, an increase in strength with time upon water immersion. Does Professor Birchall know if the water sensitivity is in any way related to the organic polymer?

Second, it appears that Professor Birchall has achieved some measure of success with incorporation of filler in the MDF cements. What was the particle size of the silicon carbide particles shown in his one microstructural slide?

J. D. BIRCHALL. We find that the water sensitivity (fall in modulus and strength on immersion in water, reversed on drying) is related to the proportion of interparticle bonding that is inorganic as against that via the organic polymer. This can be varied and as inorganic bonding becomes predominant, the sensitivity to water decreases as illustrated by the behaviour of the compact.

The median size of the silicon carbide filler was 30 μm.

J. BENSTED (*Blue Circle Technical Research Division, Greenhithe, U.K.*) The low degree of hydration of the cement in these materials creates a potential unsoundness problem if hydration continues. How durable are these materials in a wet environment?

J. D. Birchall. Extensive studies are being made, at present, into durability and the effect of exposure to water. Two aspects are being considered: first, the effect of weathering (both natural and accelerated) and second, the effect of prolonged immersion. The compact nature of the material is such that the rate of penetration of water is some 10^6 times lower than for normal cement pastes, so that in outside exposure the material never becomes saturated. Samples exposed for over 12 months have shown only slight reduction in properties. Total immersion can result in a fall in both modulus and flexural strength, which is reversible on drying and with modulus creeping higher than the original dry value as wet–dry cycling is continued. In recently prepared materials, these changes on immersion in water have been stabilized. Obviously, much more work will be needed to be done on long-term durability. It should be emphasized that considerable modification and improvement to MDF cements seems possible as experience and knowledge grow.

Phil. Trans. R. Soc. Lond. A **310**, 43–51 (1983) [43]
Printed in Great Britain

Microstructure and chemistry of unhydrated cements

By G. R. Long
Blue Circle Technical, Research Division, Greenhithe, Kent, DA9 9JQ, U.K.

Cement clinker microstructure is a major factor in determining resultant cement properties and may be modified by changes in clinker chemistry, burning régime and manufacturing processes.

1. Introduction

Optical microscopy is a very powerful technique when applied to the study of Portland cement clinker. The technique employed is essentially the polished section method developed by Tavasci (1934) although thin sections and powder mounts are used to measure optical properties.

In contrast to other techniques such as X-ray diffraction or infrared spectroscopy, which can only present average qualitative and quantitative phase analysis, the polished section technique yields specific information on crystal size, distribution of the phases and heterogeneity as well as thermal history. The information obtained has been successfully applied in recent years to assessing, modifying and improving the quality of Portland cement.

The technique can also be usefully applied to high-alumina cement clinker, although this appears to have been restricted solely to identifying and quantifying the minerals present with little information published relating microstructure to quality and hydraulic activity.

This paper discusses the interrelations between Portland cement clinker chemistry and microstructure that are known to affect cement quality in terms of strength growth pattern, workability, flowability and volume stability. The understanding gained has enabled cements with specific properties to be developed and these are briefly described.

Phase identification in high-alumina cement clinker is also briefly discussed.

2. Portland cement clinker

Portland cements are based on the C–A–F–S system; the four main phases present are C_3S, C_2S, C_3A and a ferrite phase approximating to C_4AF. C_3S (or alite) has been shown to be the main contributor to strength development, with C_2S (or belite) contributing more to late strength. The C_3A and ferrite compose the bulk of the liquid at burning temperature and although they do not contribute significantly to cement strength they play an essential fluxing role in the sintering reactions that take place in the rotary kiln.

Comparison of the quality of cements of similar chemical composition manufactured under controlled laboratory conditions indicate that although the silicate content is the main factor governing cement strength, other factors related to the clinker microstructure influence cement quality (see table 1).

To define the microstructure, polished sections of whole clinker nodules are etched with hydrofluoric acid vapour. This etch produces a straw brown colour on the angular alite crystals, while the rounded belite displays a variety of colours from deep blue to dark red. The aluminate is etched grey leaving the brightly reflecting ferrite phase apparently unattacked. Other minor constituents such as free lime, periclase, alkali modified aluminate, alkali sulphate, calcium langbeinite, anhydrite, oldhamite, spurrite, calcium sulphosilicate, $C_{12}A_7$ and $KFeS_2$ can also be identified with this etching technique.

(a) *Chemical composition*

The microstructure of the clinker, particularly the relative proportions of the phases present, depends largely, of course, on its chemical composition In the U.K. the chemical composition is defined by three ratios: the lime saturation factor (l.s.f.), silica ratio (s.r.) and alumina ratio (a.r.). These ratios can be related to the degree of burning required for adequate combination and are effective guides to the potential quantities of alite, belite, aluminate, and ferrite in the clinker. An increase in the l.s.f., which increases the alite content at the expense of the belite, and other factors remaining constant, will require harder burning to reduce the level of uncombined lime to a satisfactory level.

The s.r. governs the ratio of silicates to aluminate and ferrite; the higher the s.r. the more alite and belite but the lower the clinker liquid content at a given temperature, and consequently the higher the burning temperature required to attain combination.

The a.r., which governs the ratio of aluminate to ferrite, also indicates both the quantity of the initial liquid phase formed during firing and the relative viscosity of the liquid phase at the sintering temperature.

Increases in l.s.f., s.r. and a.r. make combination more difficult to achieve, therefore these chemical factors dictate the degree of burning the clinker requires to obtain adequate combination. Another important factor, however, is the distribution of the phases, which can be influenced by the nature of the raw materials and the manufacturing process. In particular, the preparation of the raw feed (how well it is blended and how finely it is ground), the burning process (the heating régime and atmosphere), the fuel employed and the cooling rate have all been found to influence the distribution of the clinker phases.

(b) *Alite size*

Experimental work relating alite size to the degree of burning and cement quality showed a direct relation between alite size and firing temperature and an inverse relation between alite size and cement strength.

This is shown in table 1, where it can be seen that at very small alite size the alite content is also a limiting factor. The latter relation has been used by Ono (1973) as part of a rapid microscopic method of predicting the 28 day concrete strength. Unfortunately there is no absolute firing temperature to obtain the optimum alite size and content since changes in l.s.f., s.r. and a.r. will alter the required firing temperature and affect crystal size development.

(c) *Blending of materials*

The heterogeneity arising from coarse raw materials can affect alite crystal size in two ways. First, the coarser particles will require a higher firing temperature to attain combination, which will promote alite crystal growth, and second, the presence of coarse silica will also induce large alite crystals to form.

By examining the reactions that occur as a raw material is clinkered it can be seen that small particles of silica and clay produce small belite, while large silica particles initially form a silica rich melt at low temperature. This is sucked into the adjacent porous material formed by decarbonation and as the temperature increases reacts to form a pore centered cluster of densely packed belite crystals. Belite formed in this way does not usually contain the ions needed to stabilize it in the β form and it readily inverts to the γ form on cooling, the associated volume increase breaks up the clinker. With time or temperature or both, lime is transported to belite, converting it to alite, the size of the alite formed being related to the size of the belite from which it forms. Thus coarse silica will promote the formation of large densely packed alite crystals, and usually these contain small belite inclusions.

Table 1. Relation between alite crystal size and quality

(Experimental clinkers of identical composition burnt at different temperatures.)

	burn 1	burn 2	burn 3	burn 4
burning temperature/°C	1350	1400	1450	1500
	clinker composition measured microscopically (percentage by mass)			
alite	63	70	73	77
belite	22	15	8	5
celite (flux)	10	12	17	17
free CaO	5	3	2	1
average alite crystal size/μm	13	15	19	24
	compressive strengths† of 100 mm concrete cubes/(N mm^{-2})			
3 days	16.5	18.5	17.5	16.5
7 days	24.0	29.5	27.0	25.5
28 days	33.0	44.5	42.5	40.5

† Cement specific surface area 330 $m^2\ kg^{-1}$.

There are no low temperature melts formed at the high lime end of the phase system and clusters of free lime are formed from coarse limestone particles. However, these particles lose a considerable part of their mass during decarbonation and the porous structure that results absorbs the clinker liquid phase thereby aiding assimilation of the lime. Localized concentrations of periclase, possibly associated with free lime, may occur from coarse particles of dolomite. Coarse shale particles also form belite clusters but in this instance the presence of alumina and iron results in the belite having associated aluminate and ferrite phases within the cluster.

Where the kiln is fired with coal, poor distribution of the coal ash may form similar clusters of belite dispersed in aluminate and ferrite phases. However, the size of the belite clusters formed can be several orders larger and, depending on the cause of the poor ash distribution, may occur either as peripheral layers or veins in clinker nodules, or even as cores to clinker nodules. The presence of residual carbon in the coal ash can induce an associated localized reducing environment in the clinker, which, as described later, adversely modifies the cement properties.

Poor blending of raw materials (rare in modern cement plants where sophisticated measures of control are employed) produces intranodular heterogeneity, such that the overlimed and underlimed clinker nodules that are formed have to be fired at excessively high temperatures to obtain the required combination. In fact all of the above heterogeneity effects require higher temperatures to achieve combination, with the result that alite size increases and cement quality decreases.

(*d*) *Cooling rate*

Since the aluminate and ferrite phases are liquid at firing temperature, their crystal size development can be related to the rate of cooling that the clinker experiences. Very rapid cooling, achieved by water quenching for example, may retain the liquid as a glassy phase, whereas slow cooling produces large aluminate and ferrite crystals.

A typical clinker contains about 25% liquid at the burning temperature, the crystalline phases being principally C_3S and C_2S. Under moderately fast cooling conditions the liquid phase solidifies without any significant interaction with the silicate minerals, i.e. non-equilibrium conditions prevail. But under slow cooling conditions, equilibrium is more nearly established, and at high a.r's results in a reduction of C_3S content. This is observed as erosion of the alite crystals leaving a layer of belite on the surface. Simultaneously, silica crystallizes out of the liquid phase as small 'pinhead' crystals of incipient belite. At a.r's above 2.5 slow cooling can result in a reduction of the 28 day concrete strength by up to 10%.

If during slow cooling the temperature of the clinker is maintained at about 1200 °C for any appreciable time, the alite tends to degrade, producing intimately mixed belite and free lime pseudomorphic after alite. This instability is accentuated when the clinker has been fired in a reducing environment, as discussed in §2(*g*).

(*e*) *Magnesia*

It is known that the presence of periclase may cause long term concrete durability problems. In this respect both size and quantity of periclase crystals have been reported to be important, and are governed by the firing temperature and cooling rate that the clinker experiences. With high temperature firing (over 1500 °C) the MgO goes into solution in the clinker liquid phase, and upon rapid cooling or water quenching, much of the MgO remains in solid solution and only a small quantity of small periclase crystals form. Upon slow cooling from high temperatures, around 1.5% (by mass) MgO is retained in solid solution while any quantity over this will form large periclase crystals. At lower firing temperatures (under 1450 °C) MgO is not so readily taken into solution and small periclase crystals usually form, regardless of clinker cooling rate.

Our work indicates that both free lime and magnesia have to be considered for concrete stability, particularly as determined by the ASTM C151 autoclave expansion test.

(*f*) *Alkalis and sulphates*

Alkali sulphates are among the last compounds to crystallize and occur in the clinker pore structure. Alkali sulphates are etched black with hydrofluoric acid vapour and characteristically inhibit the etching of the adjacent crystals, producing an unetched halo.

Both sodium sulphate and potassium sulphate are known to modify the concrete strength growth pattern; potassium sulphate can also promote air setting of the cement by reacting with gypsum to form syngenite.

When there is an excess of sulphate over the equivalent alkalis, the double salt of potassium and calcium, known as calcium langbeinite, is found in the interstitial phase. This compound also crystallizes into available pore space but causes less problems with air setting than potassium sulphate. Both alkali sulphate and calcium langbeinite form low temperature melts and can play a useful role in the modern dry process of manufacturing clinker by promoting agglomeration and nodulization of the feed.

When the clinker contains an excess of alkali over the equivalent sulphate, the prismatic alkali-modified aluminate phase and the alkali-modified belite form may be found. If sufficient excess alkalis are present they may stabilize belite to the extent that it will no longer react with lime to form alite, the result being an intimate mixture of belite and free lime.

(*g*) *Reducing environment*

When a normal Portland cement clinker is fired in a reducing environment the iron- and sulphur-containing phases are affected. Fe^{III} is initially reduced to Fe^{II}, and upon severe reduction to metallic Fe, with the loss of ferrite phase and an increase in aluminate content, while the clinker colour changes from grey to brown. FeO that forms goes into solid solution with lime, or substitutes for CaO in compounds, thereby increasing the overall basicity of the mix. The effective l.s.f., s.r. and a.r. of the clinker are raised, making combination more difficult; the higher firing temperature required for combination increases alite crystal size and cement quality is depressed. Alite that forms in a reducing environment and contains FeO as a lime replacement is more susceptible to degrading to belite and free lime when slowly cooled. Reoxidation of the Fe^{II} in alite at low temperature (below 1300 °C) causes the Fe^{III} formed to be thrown out of the alite crystal structure. The Fe_2O_3 reacts with the CaO in the alite and an intimate mixture of dicalcium ferrite and belite forms along the alite crystal lattice planes.

The effect of a reducing environment upon the sulphate-containing phases depends on the level of reduction. At a moderate level SO_2 is evolved from sulphates and is swept away by the kiln gases. What happens to the alkalis at this stage is not yet clear, but if quickly cooled the clinker becomes deliquescent when exposed to moisture. When such clinker is ground to a cement, the water pick-up causes surface hydration of particles and the cement will have poor flow properties. The cement will also have an increased water demand to produce a workable concrete, which will depress concrete strength. With slow cooling, alkali modified aluminate forms, which depresses the late concrete strength development.

Sulphides form with more severe reduction of the sulphates with the formation of oldhamite and $KFeS_2$. The effect of these compounds on cement quality is uncertain because of the overlying effect of all the other changes resulting from reduction.

3. Special types of Portland cement

(*a*) *White Portland cements*

White cements are produced by carefully selecting raw materials low in iron and other transition elements, and firing in a controlled reducing environment such that the small quantities of iron oxide present are converted to Fe^{II}, thus avoiding the formation of the strongly coloured ferrite phase. The clinker is water quenched from burning temperature to prevent reoxidation.

(*b*) *Sulphate-resisting Portland cement*

Sulphate-resisting clinker requires a minimal C_3A content and is produced from high iron materials to ensure that ferrite is formed in preference to the aluminate phase. The highly mobile iron rich flux promotes alite crystal growth and, at the same fineness, strength development is less than with ordinary Portland cement. The purpose of sulphate-resisting Portland cement is to produce concrete that will resist chemical attack; this requires an impermeable concrete made according to BRE Digest 250 and hence strength development is not so important.

(*c*) *Expansive cements*

Various types of expansive cement have been developed and are generally based upon the expansive formation of ettringite, brucite, or portlandite.

Cements of high free lime content that are used to produce expansion through portlandite formation unfortunately suffer from premature aeration of the free lime, thereby changing the expansive properties. To overcome the effects of aeration several processes involving the encapsulation of free lime in a melt have been developed, the product being blended with Portland cement.

Pollitt & Brown (1975) patented a novel method of encapsulating free lime within individual alite crystals, obtained by the use of calcium sulphate as the predominant clinker liquid phase. It was found that the presence of calcium sulphate stabilized belite, preventing it reacting with free lime. On firing at temperatures in excess of 1500 °C desulphation occurred, the belite was destabilized and reacted readily with adjacent free lime crystals to form alite, which contained free lime inclusions. The hydration of the free lime is delayed until the surrounding alite hydrates, and hence the expansion occurs without loss of concrete strength, unlike other expansive systems.

(*d*) *Mineralized cements*

In an endeavour to lower the firing temperature of the expansive clinker described above, calcium fluoride was added to the raw materials. The calcium fluoride inhibited the belite stabilization and the alite formed no longer contained free lime inclusions. Thus by firing relatively pure limestone and silica with gypsum and calcium fluoride at moderate temperatures (*ca.* 1400 °C), a clinker rich in alite and anhydrite was produced. This was shown to produce a high late strength cement.

Potassium sulphate was also incorporated into the raw materials with the fluoride and gypsum to change the slowly soluble anhydrite in the clinker into the more readily soluble calcium langbeinite, which provides sufficient calcium and sulphate ions for control of the initial hydration reactions. The calcium langbeinite modified the cement strength growth pattern to give both high early and high late strength. Microprobe analysis of the trigonal alite formed showed it to contain more potassium, sulphur and aluminium than the trigonal or monoclinic types of alite found in normal production cement. This cement is discussed by Dr Moir in this symposium.

4. Aluminate cements

(*a*) *Cements based on* $C_{12}A_7$

Phase equilibria precludes the formation of $C_{12}A_7$ and C_3S in the same composition. Therefore, to utilize the quick setting and early strength development of $C_{12}A_7$ in Portland cement compositions, it is necessary to make two separate clinkers and intergrind them.

A cement based on $C_{12}A_7$ was developed by Murray & Brown (1975) as a quick setting cement for binding coal contaminated aggregates at high w/c ratios, in connexion with the construction of 'Gateside Packs' in the Longwall coal mining technique used in the U.K.

A special clinker containing mainly C_2S and $C_{12}A_7$ is made in a rotary kiln from a mixture of limestone bauxite and clay. Low iron materials are selected to minimize the quantity of C_4AF and thus prevent the depletion of $C_{12}A_7$. Careful control of the raw mix is required to avoid the formation of C_3A or CA. To obtain the required strength growth characteristics, the special clinker is ground with Portland cement, the $C_{12}A_7$ producing early strength, the C_3S and C_2S ensuring a continued strength growth.

Investigations into the effect of fluorides on the C–A–S system showed that fluorine substituted for oxygen in $C_{12}A_7$ to form a modified version $C_{11}A_7 \cdot CaF_2$. It was also found that, in the presence of sufficient quantities of fluorine, the equilibrium composition could be changed from the C_3S–C_2S–C_3A compatibility triangle to the C_3S–C_2S–$C_{11}A_7 \cdot CaF_2$–CaF_2 compatibility volume.

By using fluorine rich compositions, quick setting cements of high early strength composed mainly of C_3S and $C_{11}A_7 \cdot CaF_2$ were produced by Greening (1971). Setting times as low as 10 min and significant strength development after 1 h were obtained, while final strengths were comparable to those of normal Portland cements. This cement required careful control of firing temperature. The temperature had to be sufficient to ensure adequate combination but not so high as to cause loss of fluorine such that C_2S and C_3A would form instead of the required $C_{11}A_7 \cdot CaF_2$ and C_3S.

(*b*) *High alumina cement*

High alumina cement is based around the CA region in the C–A–F–S system. As the main component of the clinker, CA produces normal setting but rapid strength development. CA occurs as short prismatic crystals and in polished sections it can be etched a blue–brown colour when immersed in boiling 10% (by mass) NaOH solution for 20 s. $C_{12}A_7$, which is usually restricted in quantity since it imparts undesirably rapid setting properties to the cement, can be identified by etching with boiling 1% (by mass) borax solution for 30 s.

Pleochroite is formed in the reduced regions of the clinker. It consumes alumina that would otherwise form CA, and if present in any quantity would cause a marked reduction in strength development. The composition of pleochroite is still in dispute, but it approximates to $C_{22}A_{13}F_3S_4$. Pleochroite is also etched by boiling 1% borax solution but can be readily identified in polished section by its fibrous or needle habit. C_2AS (or more correctly melilite) can be coloured with hydrofluoric acid vapour etch or an alcoholic nitric acid etch. It has poor hydraulicity and, as it consumes alumina that would otherwise form CA, its presence lowers the early strength development. Therefore the silica content of the cement is usually controlled to limit the C_2AS quantity.

Both HF vapour and alcoholic nitric acid etch C_2S although a more specific etch is 10% (by mass) $MgSO_4$ solution at 50 °C. The quantity of C_2S is usually low and it makes little contribution to strength development. The ferrite phase approximating to C_4AF also makes little contribution to strength and can be identified by its high reflectivity.

CA_2 is more usually formed in white high alumina cement, its high alumina content making it more applicable for refractory purposes than CA, although its strength development is more moderate. Glass can be etched with HF vapour and alcoholic nitric acid, but does not significantly contribute to strength development.

A small amount of perovskite may be formed from the TiO_2 present. In the fusion process of manufacturing high alumina clinker, the raw materials are fired in a reducing environment to aid melting. When the molten material is cast, the sudden cooling as the melt strikes the mould produces a layer of glass at the base of the cast. The rest of the cast undergoes variable cooling with the top layer quickly cooled and the centre slowly cooled. Variable reoxidation can occur as the melt is run into the mould and as it cools. These effects produce both a variation in crystal size, and a heterogeneous distribution of compounds throughout the cast.

The clinker cooling rate is reported to modify the strength growth properties. Very quick cooling markedly reduces the rate of strength development whereas very slow cooling leads to a growth in size of the CA crystals, and promotes both quicker setting and a higher initial strength.

5. Conclusions

The application of the polished section technique to cement clinkers enables an insight to be gained both into their potential properties when ground to cement and also into the process variables that affect cement quality. The technique is particularly useful in defining problems, thereby enabling remedial action to be taken. This particularly applies to phenomena such as heterogeneity, reducing conditions etc.

The author wishes to thank Blue Circle Industries p.l.c. for their permission to publish this work, and Mr B. L. Morton and Dr G. K. Moir for their helpful discussions.

References

Greening, N. R., Copeland, L. E. & Verbeck, G. J. 1971 *U.S. Patent* no. 3628973.
Murray, R. J. & Brown, A. W. 1975 *British Patent* no. 1387075.
Ono, Y. 1973 *Microscopic analysis of clinkers.* Central Research Laboratory, Onoda Cement Co.
Pollitt, H. W. W. & Brown, A. W. 1975 *British Patent* no. 1406654.
Tavasci, B. 1934 Research into the Constitution of Portland Cement Clinker. *G. Chim. ind. appl.* **16**, 538–550.

Discussion

S. Sprung (*Research Institute of the German Cement Industry, Düsseldorf F.R.G.*). Does the different velocity of cooling of clinker influence the properties of cement?

G. R. Long. The cooling that affects microstructure and cement properties is largely that occurring in the time that the clinker takes to pass between the burning zone and nose ring in the kiln, i.e. above 1200 °C. Modern coolers of all types cool very rapidly from that point on with minimal effect upon microstructure. Although the design of planetary coolers usually requires the burning zone to be away from the kiln exit, a moderately fast cooling rate can still be obtained in the kiln and resultant cement strengths will be comparable to those obtained with other types of coolers.

R. W. Davidge (*A.E.R.E. Harwell, U.K.*). Could Mr Long please define the mechanism whereby fine magnesia causes long term degradation of cement? A fine dispersion of MgO particles (few micrometres) could give highly localized microcracking and hence a toughening effect as described in my paper on ceramics (this symposium).

G. R. Long. Periclase present in Portland cement is very tardy in hydrating and tends to remain as anhydrous grains after concrete made from the cement has set and hardened.

Subsequent ingress of water can expansively hydrate the periclase, thereby inducing disruptive cracking in the concrete. The occurrence of such cracking will depend upon both the size and quantity of the periclase crystals present.

D. Pearson (*A.E.R.E. Harwell, U.K.*). Are polymorphic modifications of the cement clinker constituents an important factor in the development of strength of cements?

G. R. Long. Reflected light microscopy does not lend itself to the ready identification of individual polymorphs unless considerable structural differences exist, such as between the β and γ polymorphs of C_2S. We would normally use X-ray diffraction techniques to detect the polymorphism of any constituent.

The increase in cement strengths mentioned by Dr Pomeroy was not related to changes in polymorphism of the clinker minerals. Alite forms a number of polymorphs, which are stabilized by various minor elements. As reported in Dr Moir's paper to this symposium, superior strength development properties are associated with rhombohedral alite. However, the levels and nature of the minor elements responsible for stabilization of a particular polymorph may have a much greater influence on hydraulic activity and strength properties than the actual polymorphic form itself.

Phil. Trans. R. Soc. Lond. A **310**, 53–66 (1983) [53]
Printed in Great Britain

New developments in understanding the chemistry of cement hydration

By D. D. Double
Department of Metallurgy and Science of Materials, University of Oxford, England, U.K.

[Plate 1]

Portland cement is based on calcium silicates with lesser amounts of calcium aluminates. Its reaction with water, which is responsible for strength development, is a complex process involving the precipitation of hydration products in colloidal gel and crystalline forms. This paper describes the microstructural development of the hydrates the structure of colloidal C–S–H gel, which is the main hydration product, the kinetics of hydration and the aqueous solution chemistry of cement pastes and pastes made with pure C_3S. Theories relating to the mechanism of cement hydration are examined and these are discussed in terms of studies of the effect of accelerating and retarding admixtures for cement.

1. Anhydrous cement constituents and hydration products

Portland cement sets and hardens by chemical reaction with water. The development of strength of a cement paste is associated with the precipitation of hydration products of very low solubility to form a cohesive matrix binding the residual cement grains into a composite solid.

The main constituents of anhydrous cement clinker are slightly impure forms of tricalcium silicate (C_3S, about 50% by mass), dicalcium silicate (C_2S, *ca.* 25%), tricalcium aluminate (C_3A, *ca.* 10%) and an aluminoferrite phase approximating to the formula C_4AF (*ca.* 10%). The hydration chemistry is dominated by the calcium silicate phases and these provide the main source of strength development in a cement paste. The aluminate phases, in particular C_3A, affect the setting behaviour but are not thought to contribute significantly to the overall strength attainment. The C_3S phase hydrates faster than C_2S and since it is also the major constituent in Portand cement it is often taken as the model system by experimentalists in studying cement chemistry.

A summarized scheme of the hydration process is shown in figure 1, illustrating the anhydrous constituents in relation to their hydration products. In spite of their difference in reactivity, the hydration products of C_3S and C_2S are essentially the same. The main product is calcium silicate hydrate (C–S–H) gel, a colloidal precipitate, which is mainly amorphous and which has a rather variable composition. This occupies about 60% (by volume) of the hydration products and is therefore the main binding agent in hardened cement. Calcium hydroxide is a by-product of the hydration of the calcium silicates and contributes a further 20% (by volume) approximately. The rest comprises various calcium sulphoaluminate and aluminoferrite hydrates, which are the products of hydration of the aluminate phases and gypsum. Gypsum (*ca.* 2%) is deliberately added to Portland cement to control the rapid hydration of C_3A, which otherwise tends to produce premature 'flash setting' of the cement. Apart from calcium hydroxide, crystalline products of hydration detectable by X-ray diffraction include ettringite ($C_3A.3C\bar{S}.32H$) and calcium monosulphoaluminate ($C_3A.C\bar{S}.12H$).

The reactions shown in figure 1 do not occur independently and there is appreciable fine-scale mixing of the hydrates, particularly in the colloidal gel product. During hydration, there is an increase in the volume of solids and this is accommodated by growth of the hydrates into the water filled spaces in the cement microstructure during the hardening process.

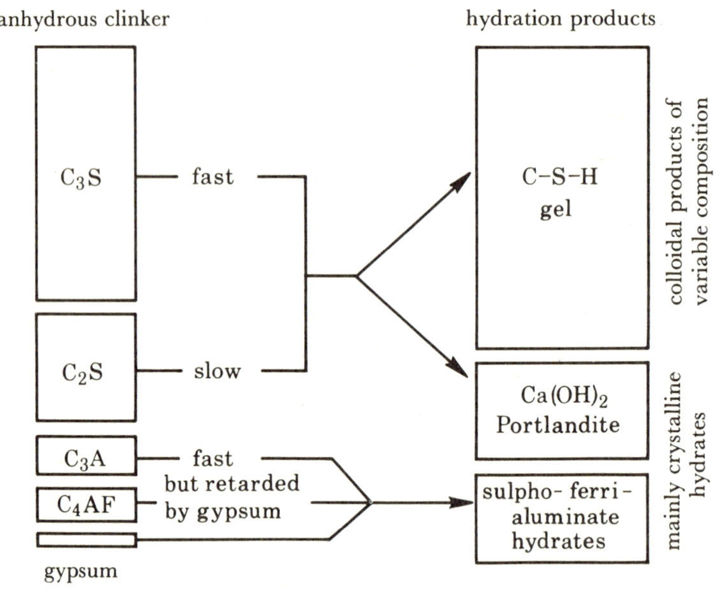

FIGURE 1. Schematic representation of the anhydrous constituents in Portland cement clinker and the products formed during hydration. The areas of the 'boxes' give the approximate volume proportions of the phases.

2. Development of microstructure during hydration

Electron microscopy has been widely used to study the formation of hydration products in cements and in samples made from pure cement compounds (Daimon *et al.* 1971; Williamson 1972; Collepardi & Marchese 1972; Lawrence & Young 1973; Diamond 1976). The main features of a hydrated cement are shown in figure 2 (plate 1). On the fractured section, the prominent morphology is a reticular structure consisting of irregular fibres growing from the surfaces of the cement grains and bridging the spaces between. This material is identified as the C–S–H gel hydration product. Large plate-like crystals of calcium hydroxide also occur in the microstructure. Similar microstructures occur in hydrated samples of pure C_3S and C_2S. The development of C–S–H gel growths in a sample of C_3S at early stages of hydration is shown in figure 3 (plate 1). In mature, fully aged cements there is much infilling and consolidation of the microstructure by fine grained C–S–H gel and by growth of crystalline hydration products. In general, the morphology of C–S–H gel is variable depending on the w/c ratio and age of the cement sample, so it is difficult to classify characteristic types.

The growth of hydration products at the surfaces of cement grains has been studied in detail by transmission electron microscopy methods (Double *et al.* 1978; Jennings & Pratt 1979; Groves 1981). These show essentially a two-stage process. On contact with water, the cement grains almost immediately become enveloped by thin coatings of colloidal gelatinous hydrates. The coatings are of the order 0.1–1.0 μm in thickness and give only weak diffuse diffraction rings against the spot patterns emanating from the underlying crystalline clinker mineral (figure 4, plate 1). At later stages, after several hours, secondary growth of C–S–H gel occurs,

FIGURE 2. Fracture surface of a hardened cement paste showing fibrillar growth of C–S–H gel around the cement grains and plate-like crystals of calcium hydroxide (w/c ratio 0.7, age 40 days). Scanning electron micrograph.

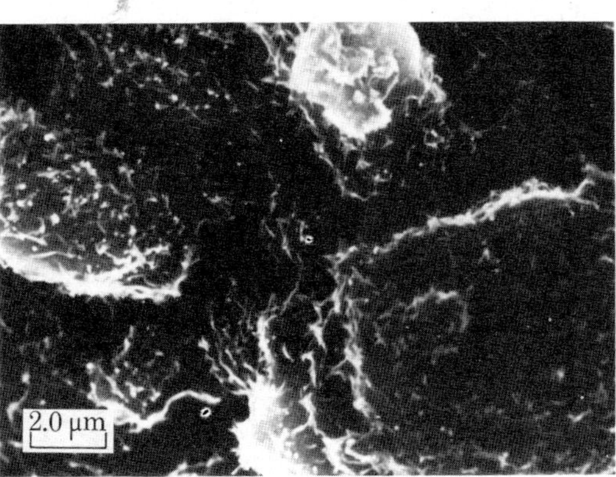

FIGURE 3. Fracture surface of a C_3S paste sample at early stages of hydration showing gel coatings around the particles and the beginning of the growth of fibrillar C–S–H gel product (w/c ratio 0.5, age 2 h). Scanning electron micrograph.

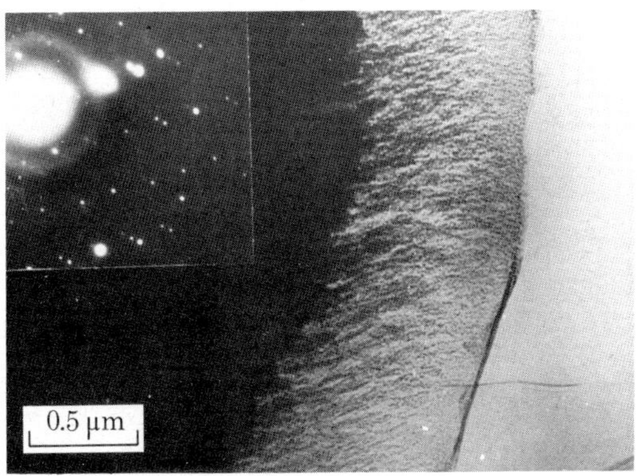

FIGURE 4. Colloidal gel coating formed on a grain of C_3S in a thinned slice of Portland cement clinker after dipping in water for 5 min. Transmission electron micrograph and selected area diffraction pattern (Groves 1981).

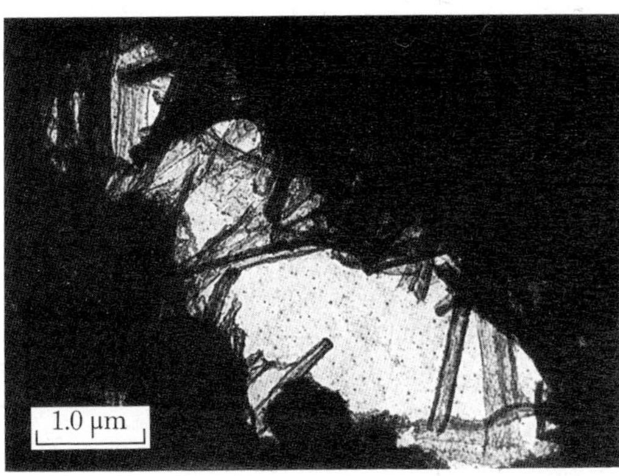

FIGURE 5. Secondary growth of C–S–H gel at the surfaces of cement grains after hydration for 21 h, the hydrate shows fibrillar and crumpled sheet type morphologies. Transmission electron micrograph.

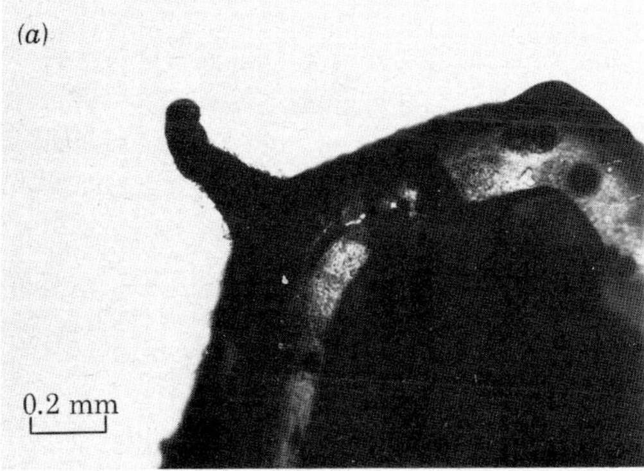

FIGURE 10. 'Silicate garden' obtained by immersing a grain of sodium silicate in dilute aqueous calcium nitrate solution. (*a*) Formation of colloidal C–S–H gel coating around the dissolving grain. (*b*) Tubular growths formed after osmotic rupture of the gel coating. Optical micrographs.

(*Facing p.* 54)

sometimes in fibrillar form and sometimes as exfoliated crumpled sheets (figure 5, plate 1). As in s.e.m. observations, t.e.m. studies show a variety of morphological forms of C–S–H gel. The observation that the fibrillar growths sometimes show hollow tubular structures (Double *et al.* 1978) is interesting, since it relates to theories of the mechanism of cement hydration. Analysis of the fibres gives compositions that include appreciable amounts of aluminium and sulphur in addition to calcium and silicate (Jennings & Pratt 1979). Amorphous tubular growths have also been reported in studies of C_3A hydrated in lime with gypsum solutions, the fibres having C–A–$\bar{S}$–H type compositions (Bailey & Chescoe 1976).

A distinction is often made, in terms of composition and structure, between 'inner' and 'outer' hydration products. This differentiates between those products precipitated within the original outline of the cement grain as it dissolves during hydration and those growing in the water-filled spaces outside. The microstructures described above constitute outer products. Inner products, adjacent to the cement grain surface, usually have a very fine grained and compact texture.

3. Composition and structure of C–S–H gel

Electron microscopy shows that the basic structure of C–S–H gel is an agglomeration of colloidal particles of rather indefinite shape but generally less than 50 nm size. This is consistent with large surface areas and very fine porosity determined by mercury intrusion and gas adsorption techniques (Brunauer *et al* 1967; Diamond 1971) and by X-ray and neutron low-angle scattering methods (Winslow & Diamond 1974; Allen *et al.* 1982). The composition of C–S–H gel is variable and probably depends on locality in the microstructure (namely, inner and outer hydration products) and on the age of the sample. Electron microanalysis studies applied to cement and C_3S samples give CaO/SiO_2 molar ratios in C–S–H gel ranging between 1.5 and 2.0, a notable feature in cement samples being the inclusion of appreciable quantities of Al, Fe, S and alkali impurities (Taylor & Roy 1980). In addition to compositional variations, structural rearrangements occur with time. Trimethylsilylation techniques have been used to study the degree of polymerization of the silicate (Taylor & Roy 1980). The conclusions are that the silicate in C–S–H gel in hydrated cement and C_3S samples is mainly dimeric at early stages of hydration ($Si_2O_7^{6-}$ units), but that there is a progressive condensation to higher polymeric forms (greater than tetramer) with ageing. Even in samples several years old, the dimeric fraction may be as much as a third of the total.

4. Chemistry of hydration

The reaction between cement compounds and water is exothermic, and calorimetric methods have proved effective in studying the sequence of hydration (Stein 1961; Tenoutasse 1969; Previte 1971; Double *et al.* 1978). Figure 6 shows the general form of the heat evolution curve obtained from a sample of C_3S.

Hydration occurs in a number of distinct stages and these are characterized by an initial very rapid evolution of heat within the first few minutes and a later more gradual peak reaching a maximum after about 8–10 h and decreasing slowly thereafter. The peaks are separated by an 'induction period' lasting between 1 and 3 h, when the rate of heat evolution drops to a very low value. The initial peak is largely due to the heat of wetting of the C_3S powder and the hydration of any free lime in the sample. An increase in temperature accelerates hydration and this is seen in the calorimetric curve as a shortening of the induction period and an increase in height of the second peak, which is brought to an earlier time interval. The same form of heat evolution is shown by cement, although modified by additional contributions from the other constituents, notably by the hydration of C_3A and at later stages by the hydration of C_2S. In terms of the microstructural developments discussed previously, the initial peak corre-

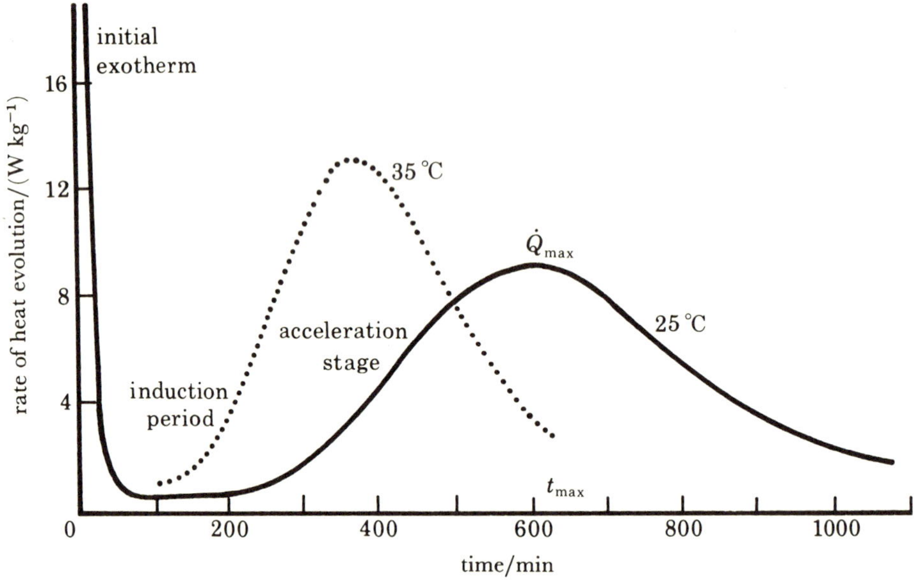

FIGURE 6. General form of the heat evolution curve obtained during the hydration of C_3S. The dotted line shows the accelerating effect of increasing the temperature of hydration.

sponds to the formation of gelatinous coatings of hydration products around the cement grains. The later accelerational period corresponds to the beginning of the secondary growth of C–S–H gel hydration products. In practical terms, the accelerational stage is important since it covers the period of setting of the cement paste under normal conditions.

It is relevant to consider the calorimetric curves in relation to changes in the aqueous solution chemistry of a cement paste. Techniques involve extracting and filtering the liquid phase from a cement or C_3S paste at various stages of hydration and analysing for dissolved species (Lawrence 1966; Tadros *et al.* 1976; Young *et al.* 1977; Barret & Ménétrier 1980; Thomas & Double 1981; Thomas *et al.* 1981). Results tend to vary depending on the methods used, but the curves shown in figure 7*a*, *b* are generally typical.

In the simpler case of a pure C_3S paste (figure 7*a*), there is initially a very rapid rise in the calcium and hydroxyl ion levels within a few minutes of mixing with water. This is followed by a more gradual increase, which reaches a maximum after about 3–5 h. The solubility limit of calcium hydroxide is exceeded within about 20–30 min and the final peak represents a supersaturation by a factor between 1.5 and 2. A significant feature is that during this period the concentration of silicate in solution is almost negligibly small, generally less than $2/10^6$. In relation to the calorimetric curves, two points are worth noting. First, that the calcium and hydroxyl ion concentrations are continuously rising during the induction period. Second, that the supersaturation peak corresponds approximately to the end of the induction period and the stage of accelerated reaction.

In a cement paste (figure 7*b*), the situation is more complex but the underlying features are

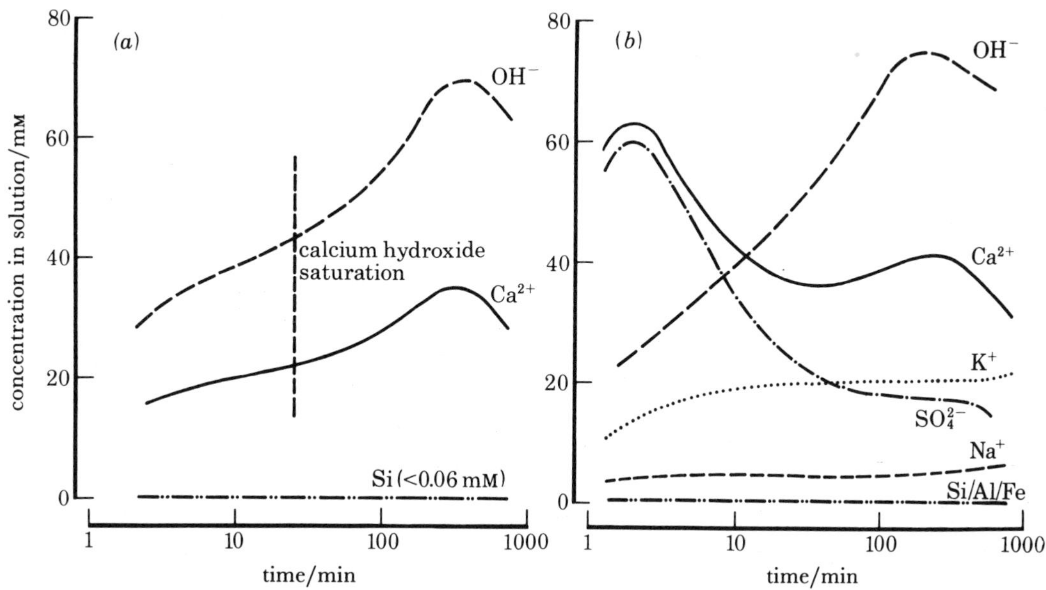

FIGURE 7. Typical concentration against time curves obtained by chemical analysis of solutions filtered from (*a*) a C_3S paste sample and (*b*) Portland cement paste (w/c ratio = 2).

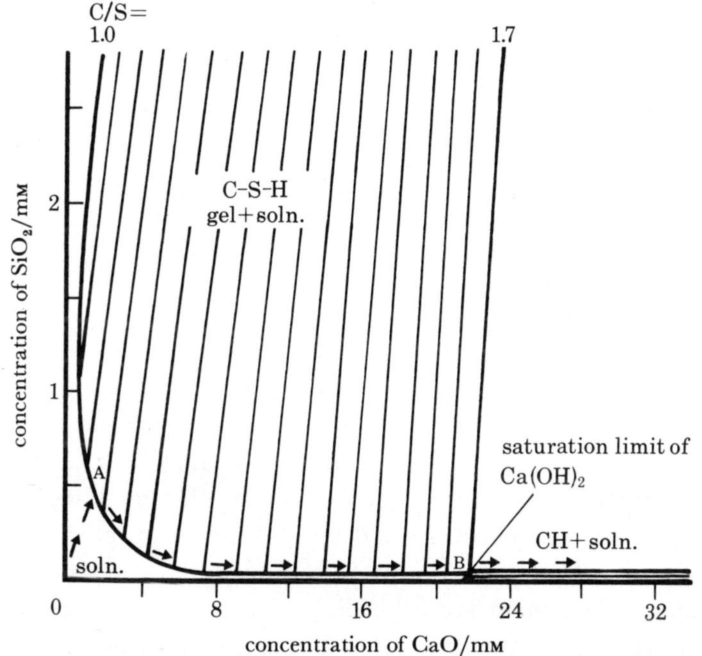

FIGURE 8. General form of the solubility diagram for the ternary $CaO–SiO_2–H_2O$ system at room temperature (based on Flint & Wells 1934). The arrows indicate the expected changes in lime and silica concentrations in solution during the hydration of C_3S.

similar. Solutions show rising concentrations of calcium and hydroxyl ions reaching a supersaturation peak, and silicon concentrations that remain extremely small throughout. Aluminum and iron are also almost absent from solution (*ca.* $2/10^6$ or less). Superimposed on this are increasing concentrations of sodium and potassium from the alkalis in cement and an early peak in the concentrations of calcium and sulphate. The latter may be attributed to the dissolution of gypsum in the cement and its combination with the aluminate phases to precipitate sulpho-aluminate hydrates.

The calcium silicon ratios in solution from cement and C_3S pastes are generally in the range $(10^2–10^3):1$. It has been shown that these concentrations may be correlated with data from experimentally determined solubility limits in the ternary $CaO–SiO_2–H_2O$ system (Thomas & Double 1981). The solubility diagram shown in figure 8 is based on the work of Flint & Wells (1934), but is drawn on an exaggerated scale for clarity (the solubility of silica in lime solution is very much smaller than shown, generally less than 0.05 mM).

The solubility diagram illustrates two points. The first is the insolubilization of silica in the presence of lime by precipitation as C–S–H gel. The second is the variability in the CaO/SiO_2 ratio of the C–S–H gel depending on the composition of the solution with which it is in contact (as indicated by the changing slopes of the tie-lines in the two phase C–S–H gel + solution region). Solution compositions in C_3S pastes generally lie along the line AB, the solubility limit of C–S–H gel, with lime concentrations increasing with time. The point B, which is an invariant point where both the solids C–S–H gel + $Ca(OH)_2$ are in equilibrium with saturated solution, is in practice exceeded as the solution approaches the supersaturation peak. Under normal circumstances, the initial hydrolysis of C_3S probably does not involve congruent dissolution. In any case the solubility limit of C–S–H gel is reached very soon after the contact of C_3S with water, and this is associated with the almost immediate precipitation of C–S–H gel at the surfaces of the grains. The rising concentration of lime in solution is a consequence of the precipitation of C–S–H gel with a lower CaO/SiO_2 ratio (ranging between 1.5 and 1.7 according to the solubility diagram) than the surrounding solution.

5. Mechanism of cement hydration

Questions relating to the mechanism of cement hydration are the subject of much debate in the literature (Skalny & Young 1980). The situation is complicated by the possibility that different rate controlling factors may operate at various stages of hydration. As far as early hydration is concerned, one can identify at least three possible rate determining processes: (*a*) the hydrolysis of the anhydrous clinker phases, (*b*) the nucleation and growth of the hydration products and (*c*) the diffusion of water and solute species through coatings of hydration products around the cement grains. That stage (*a*) may be important is indicated by reports that the reactivity of C_3S is influenced by polymorphic modification and by defects introduced by prior grinding or quenching and by u.v. excitation of the clinker (Fierens & Verhaegen 1976; Maycock *et al.* 1978; Aldous 1983).

In general terms, models for cement hydration divide into two categories: 'delayed nucleation theories' and 'protective coating theories'. The former mainly concentrate on the delayed nucleation of calcium hydroxide, associating the end of the induction period with the supersaturation peak in solution (Tadros *et al.* 1976; Young *et al.* 1977; Dent Glasser *et al.* 1978). The nucleation of calcium hydroxide is seen as being poisoned by the adsorption of small amounts of silicate from solution. The increasing concentration of lime in solution is then seen as progressively inhibiting the hydration of the cement silicates (induction period), a situation that is relieved only when $Ca(OH)_2$ eventually nucleates and grows. Evidence contrary to these ideas comes from experiments that show that the hydration of cement or C_3S is not significantly accelerated either by hydrating in already saturated lime solution or by prior 'seeding' of the paste sample by crystalline calcium hydroxide (Odler & Dorr 1979). Also, the presence of crystalline calcium hydroxide has sometimes been detected long before the supersaturation peak and the end of the induction period (Slegers & Rouxhet 1977). Alternative theories that invoke delayed nucleation of C–S–H gel as the rate controlling factor (Fierens & Verhaegen 1976; Odler & Dörr 1979) do not seem compatible with studies of the aqueous solution chemistry or electron microscopy observations that show almost immediate precipitation of C–S–H gel when cement silicate grains contact water.

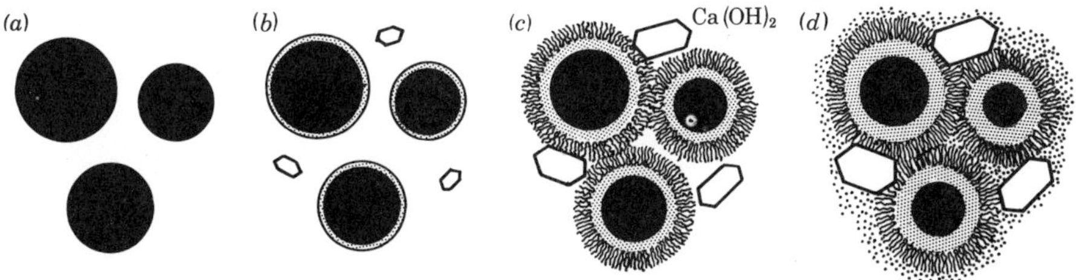

FIGURE 9. Schematic representation of the sequence of hydration of cement. (*a*) Cement grains in water. (*b*) Formation of protective colloidal coatings of C–S–H gel around the cement grains. (*c*) Rupture of the protective coatings followed by secondary growth of C–S–H gel. (*d*) Later infilling of the microstructure by fine grained C–S–H gel and by growth of crystalline calcium hydroxide.

Protective coating theories view the growth of calcium hydroxide as the effect of accelerated hydration rather than the cause. The proposed sequence of events is schematically illustrated in figure 9. The rapid precipitation of C–S–H coatings around the cement grains provides protective barriers that inhibit the access of water and retard hydration during the induction period. The end of the induction period occurs when the coatings are disrupted, thus leading to accelerated hydration and growth of secondary C–S–H gel hydration products. The acceleration period is followed by a gradually decreasing rate of reaction and this is attributed to later infilling and accretion by hydration products. Effectively, as the layers of hydration products thicken by growth around the cement grains and as free water in the microstructure is consumed, longer range diffusional processes dominate and progressively slow down hydration.

The important question is the cause of the disruption of the initial protective coating. One explanation is that there is a spontaneous change in the chemical and physical structure of the C–S–H gel, a transition to a secondary C–S–H hydrate that is more permeable to the diffusion of water and solute species (de Jong *et al.* 1967). There is no direct experimental evidence, however, that such a phase change occurs in C–S–H gel within the induction period. An alternative proposal, based on an original suggestion by Powers (1961), is that the coatings are ruptured by osmotic pressure effects due to the selectively permeable character of the colloidal gel. This is the basis of the 'osmotic membrane' model for cement hydration (Double *et al.* 1978, 1980; Birchall *et al.* 1978, 1980). In this model the C–S–H gel coating is seen as acting as a membrane that allows inward diffusion of water to the cement grains and outward diffusion of calcium and hydroxyl ions released by hydrolysis. The membrane is, however, effectively impermeable to the outward diffusion of the larger and sterically hindered silicate ions. Preferential diffusion leads to a rising osmotic pressure within the coating sufficient eventually at the end of the induction period, to cause its rupture and exfoliation. The hydrous silicate, originally retained within the membrane coating, is extruded into outer lime-rich bulk solution where it is immediately precipitated as secondary C–S–H gel products growing from the surfaces of the original coating. The model explains the transport of solute species during hydration and precipitation (calcium and hydroxyl ion concentrations rising during the induction period, yet negligible quantities of silicate in the bulk solution) and is also consistent with the observed morphological development of the secondary C–S–H gel hydration products (in fibrillar, crumpled sheet and tubular forms).

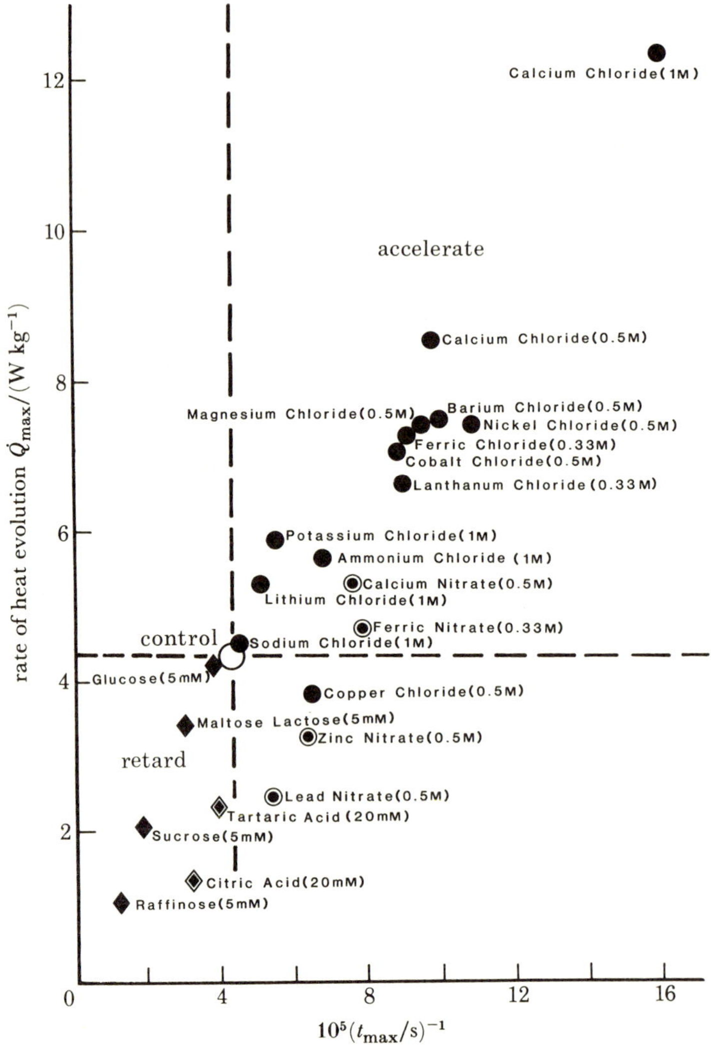

FIGURE 11. Plot of $\dot{Q}_{max}$ against $1/t_{max}$ from calorimetric heat evolution curves (see figure 6) showing the acceleration and retardation of cement hydration by a range of water soluble admixtures (w/c ratio 0.5, temperature 27 °C).

In this context, attention has been drawn to very similar growth forms, albeit on a much magnified scale, found in 'silicate gardens'. These are obtained when water soluble salt crystals with cations other than those of the alkali metals are immersed in dilute sodium silicate solution (Coatman *et al.* 1980). The analogy demonstrates very graphically the selectively permeable and osmotic characteristics of inorganic colloidal gel membranes. The effect is not confined to silicates but includes aluminates and other oxyanion species when precipitated with a metal cation as a continuous colloidal gel membrane between two solutions of differing concentrations. Figure 10*a*, *b* (plate 1) show a 'reverse' silicate garden, a sodium silicate grain immersed in a dilute solution of a calcium salt. This provides a fairly close analogy with the situation in cement. The photographs show the initial precipitation of an envelope of colloidal calcium silicate hydrate (C–S–H) gel around the dissolving silicate grain and later tubular growths developed after rupture of the coating by osmotic pressure effects.

The protective membrane model does not preclude the precipitation of hydration products within the initial gel coating. Inner products can form as the cement grains dissolve away during hydration. Since, however, the volume of hydration products exceeds that of the dissolved anhydrous material (see figure 1) not all can be accommodated in this way. Inner precipitation processes can thus be seen as an additional source of pressure within the gel coating.

6. Accelerating and retarding admixtures

Water soluble admixtures are widely used by the industry to accelerate or retard the setting of cement. Many inorganic salts have an accelerating effect on cement hydration and the most important in commercial terms has been calcium chloride. Exceptions are Zn and Pb salts and phosphates and borates, which have a retarding effect. Commercial retarders, however, are mainly based on organic admixtures such as hydroxycarboxylic acid and sugar derivatives and these can be effective at very low concentrations.

Calorimetric methods have proved effective in classifying admixtures (Previte 1971; Kantro 1975; Kondo *et al.* 1977) but comparison of results is often difficult because of the use of different parameters to measure acceleration or retardation. The system used here is based on a plot of the maximum in the heat evolution curve ($\dot{Q}_{max}$) against the reciprocal of the time to the peak maximum ($1/t_{max}$). Figure 11 shows that this separates admixtures along a roughly diagonal relation according to their accelerating (top right) or retarding (bottom left) capacity relative to the untreated control cement. At the extremes, the acceleration by calcium chloride and the retardation by citric acid and raffinose are evident in this plot. The degree of acceleration or retardation depends on the concentration of the admixture.

In inorganic salt admixtures both cation and anion can be effective to varying degrees and attempts have been made to rank cations and anions separately according to their accelerating or retarding stength. For example our calorimetric system applied to cements treated with chlorides and nitrates (at equivalent concentrations of anion, 0.5 gram-ion per litre) gives approximately the following ranking sequence for cations:

$$\tfrac{1}{2}Ca^{2+} > \tfrac{1}{2}Ni^{2+} > \tfrac{1}{2}Ba^{2+}, \tfrac{1}{2}Mg^{2+} > \tfrac{1}{3}Fe^{3+}, \tfrac{1}{2}Co^{2+} > \tfrac{1}{3}La^{3+} \gg NH_4^+, K^+ > Li^+ > Cs^+ > Na^+ >$$

←acceleration

$$\tfrac{1}{2}Cu^{2+} > \tfrac{1}{2}Zn^{2+} > \tfrac{1}{2}Pb^{2+}.$$

retardation→

Similar rankings have been found for the acceleration of the hydration of C_3S (Collepardi *et al.* 1971). For calcium salts at equivalent concentrations the work of Kantro (1975) indicates the following sequence for anions in decreasing order of acceleration of C_3S hydration:

$$Cl^-, Br^- > SCN^- > I^- > NO_3^- > ClO_4^-.$$

The importance of calcium in the ranking system is evident but other polyvalent cations can have similar accelerating strength at appropriate concentrations. Monovalent and alkali metal cations seem only weakly effective. Attempts have been made to correlate the accelerating capacity of ions in terms of their diffusivities (Kondo *et al.* 1977). Attempts to rationalize the

the ranking have generally been inconclusive. For cations, problems arise in establishing the configurations of the ionic species, which, in the highly alkaline (pH 12–13) environment of cement, will mostly be in a hydroxylated state (for example, $Ca(OH)^+$) in solution or will be immediately precipitated as hydroxide. As discussed by Kantro (1975), the relative solubilities of the hydroxides formed by the cations are probably an important factor. Studies of Pb and Zn salts show that retardation is associated with the immediate precipitation of dense and probably impermeable coatings of hydroxide gel around the cement grains (Thomas *et al.* 1981; Arliguie *et al.* 1982).

Table 1. Retarding effect of a range of organic admixtures

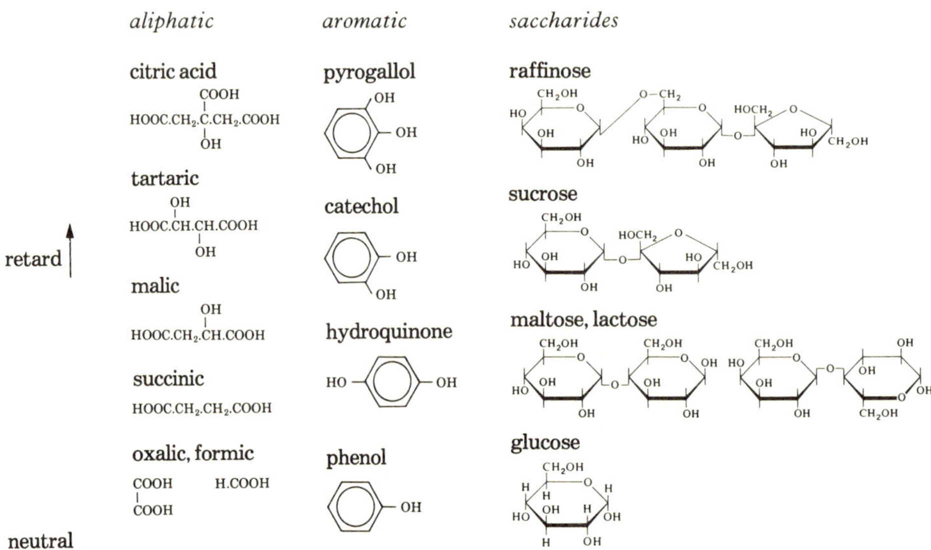

The retardation of cement by organic admixtures is primarily attributed to their effect on the hydration of the C_3S phase, but it is recognized that interactions with the aluminate phases may also be important. Explanations for the mechanism of retardation have included adsorption to the cement grain surfaces, the poisoning of nucleation and growth of the hydration products, and the precipitation of complexed species with the colloidal C–S–H gel (Young 1972). Table 1 above, based on the calorimetric ranking system, gives results for a range of aliphatic and aromatic admixtures (at 20 mM concentration) and saccharide admixtures (at 5 mM).

In the aliphatic acids there is a general increase in retardation with molecular mass and number of carboxylic groups but the important factor is the presence of α- or β-hydroxycarboxylic groups that are capable of strongly chelating a metal cation, such as calcium. In the aromatic admixtures the number and positioning of the hydroxyl groups is important and, again, this may be related to their capacity for complexing calcium. Saccharides are also capable of binding calcium (Rendleman 1966; Angyal 1980) although their chemistry in relation to cement hydration is complicated by their susceptibility to degradation by alkaline hydrolysis. The greater stability of raffinose and sucrose in this respect is significant since these are amongst the strongest retarders of cement hydration (Previte 1971). That complexation of calcium may be an important factor is demonstrated by the fact that abnormally high levels of silicon appear in solution at early stages of hydration in cements treated with EDTA, citric acid and sucrose (figure 12). Preferential sequestration of calcium by the organic molecule initially allows solubilization of the silicate, and subsequently affects the chemistry of precipitation of the C–S–H gel (Thomas & Double 1983). EDTA, which has a very strong calcium binding capability, is a relatively weak retarder compared to citric acid and sucrose. This suggests that additional factors such as the complexation of silicate (or aluminate) may also be involved, either initially as soluble complexes or finally precipitated with the C–S–H gel. Complexes of silica with citric acid and catechol are known and have been cited as the cause of promoted attack of silica glass in neutral and alkaline solutions containing dissolved organic species (Ernsberger 1959; Bacon & Raggon 1959).

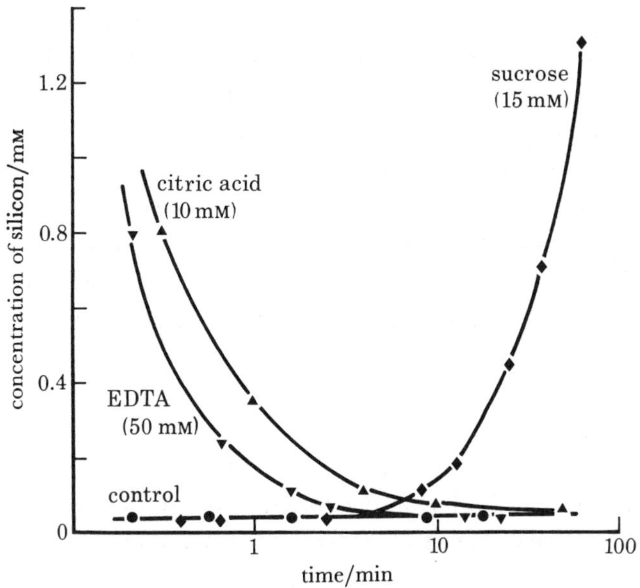

FIGURE 12. Concentrations of silicon in solutions filtered from cement pastes containing EDTA, citric acid and sucrose admixtures, compared with an untreated control sample (w/c ratio 2).

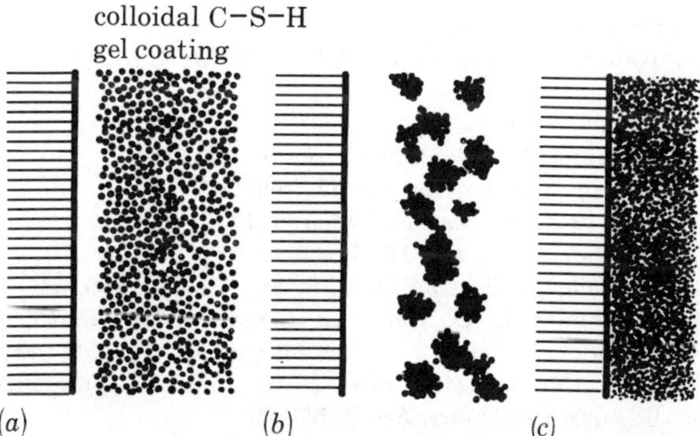

FIGURE 13. Schematic representation of possible changes in the colloidal structure of a C–S–H gel coating on a cement grain surface. (*a*) Normal protective gel coating, inhibiting hydration. (*b*) Flocculated, poorly adherent coating permitting accelerated hydration. (*c*) Dense, coagulated and strongly adherent coating causing retardation of hydration.

If, as seems likely, the hydration of cement is largely controlled by diffusion through protective colloidal coatings around the cement grains then it follows that the rate of reaction and the associated hydrate development will depend on the permeability and cohesion of the coatings. In these terms, cement hydration is normally self inhibiting. Modification of the colloidal coating to a more open flocculated structure will facilitate diffusion and accelerate hydration. Conversely, coagulation of the coating to a denser, less permeable and more adhesive layer will result in a retardation of hydration (figure 13).

In general, the principal cause of aggregation of a colloid is Van der Waals' attractive forces between the particles, whereas stability against aggregation is a consequence of repulsive interactions between similarly charged electrical double layers and layers arising from solvent–particle affinity. In aqueous systems, the pH of solution and dissolved salts strongly influences aggregation, and organic molecules can affect stability by adsorption on to the particle surfaces or by bridging between the particles. In this context, a tentative parallel can be drawn between the ranking of cations and anions for inorganic accelerating admixtures and the lyotropic series (Hoffmeister series), which relates to the efficacy of flocculation of hydrophilic colloids by dissolved electrolytes. Organic admixtures affect hydration by altering the chemistry of precipitation of C–S–H gel by preferential complexation of calcium or silicate in solution, or both, or by incorporation into the protective C–S–H gel coating. Measurements of ζ potentials for cement pastes show that the sign and magnitude of the electrical double layers are markedly affected in the presence of organic admixtures (Singh & Ojha 1980; Roy *et al.* 1980). Colloidal phenomena are important considerations in cement chemistry and clearly this is an area of research that will see further development in the future (Ottewill, this symposium).

The author, who is a Warren Research Fellow of the Royal Society, expresses appreciation for support from the Science and Engineering Research Council, from Imperial Chemical Industries p.l.c. and from Cormix Ltd in the development of this research. Thanks are due also to Professor Sir Peter Hirsch, F.R.S., for his encouragement, and to researchers at the Department of Metallurgy and Science of Materials at the University of Oxford, who have contributed to this work.

References

Aldous, R. T. H. 1983 *Cem. Concr. Res.* **13**, 89–96.
Allen, A. J., Windsor, C. G., Rainey, V., Pearson, D., Alford, N. McN. & Double, D. D. 1982 *J. Phys.* D **15**, 1817–1833.
Angyal, S. J. 1980 *Chem. Soc. Rev.* **4**, 9–18.
Arliguie, G., Ollivier, J. P. & Grandet, J. 1982 *Cem. Concr. Res.* **12**, 798–6.
Bacon, F. R. & Raggon, F. C. 1959 *J. Am. ceram. Soc.* **42**, 199–205.
Bailey, J. E. & Chescoe, D. 1979 *Proc. Br. ceram. Soc.* **28**, 165–177.
Barret, P. & Ménétrier, D. 1980 *Cem. Concr. Res.* **10**, 521–529.
Birchall, J. D., Howard, A. J. & Bailey, J. E. 1978 *Proc. R. Soc., Lond.* A **360**, 445–453.
Birchall, J. D., Howard, A. J. & Double, D. D. 1980 *Cem. Concr. Res.* **10**, 145–155.
Brunauer, S., Mikhail, R. S. & Bodor, E. 1967 *J. Colloid Interface Sci.* **24**, 451–460.
Coatman, R. D., Thomas, N. L. & Double, D. D. 1980 *J. Mater. Sci.* **15**, 2017–2026.
Collepardi, M. & Marchese, B. 1972 *Cem. Concr. Res.* **2**, 57–65.
Daimon, M., Ueda, S. & Kondo, R. 1971 *Cem. Concr. Res.* **1**, 391–401.
de Jong, J. G. M., Stein, H. M. & Stevels, J. M. 1967 *J. appl. Chem., Lond.* **17**, 246–255.
Dent Glasser, L. S., Lachowski, E. E., Mohan, K. & Taylor, H. F. W. 1978 *Cem. Concr. Res.* **8**, 733–743.
Diamond, S. 1971 *Cem. Concr. Res.* **1**, 531–545.
Diamond, S. 1976 In *Proc. Conf. Hydraulic Cement Pastes; their Structure and Properties, Sheffield*, pp. 2–30. Slough: Cement and Concrete Association
Double D. D., Hellawell, A. & Perry, S. J. 1978 *Proc. R. Soc., Lond.* A **359**, 435–451.
Double, D. D., Thomas, N. L. & Jameson, D. A. 1980 In *Proc. 7th Int. Congr. Chem. Cem., Paris*, vol. II, II.356–II.260. Paris: Éditions Septima.
Ernsberger, F. M. 1959 *J. Am. ceram. Soc.* **42**, 373–375.
Fierens, P. & Verhaegen, J. 1976 *Cem. Concr. Res.* **6**, 103–111.
Flint, E. P. & Wells, L. S. 1934 *J. Res. natn. Bur. Stand.* **12**, 751–780.
Fuji, K. & Kondo, W. 1968 In *Proc. 5th Int. Symp. Chem. Cem., Tokyo*, vol. II, paper II-93, 362–370.

Groves, G. W. 1981 *J. Mater. Sci.* **16**, 1063–1070.
Jennings, H. M. & Pratt, P. L. 1979 *Cem. Concr. Res.* **9**, 501–506.
Kantro, D. L. 1975 *A.S.T.M. J. Test. Eval.* **3**, 312–321.
Kondo, R., Daimon, M., Sakai, E. & Ushiyawa, H. 1977 *J. appl. Chem. Biotechnol.* **27**, 191–201.
Lawrence, C. D. 1966 *Highw. Res. Abstr.*, report no. 90, 378–391.
Lawrence, F. V. & Young, J. F. 1973 *Cem. Concr. Res.* **3**, 149–161.
Maycock, J. N., Skalny, J. & Kalyoncu, S. 1978 *Cem. Concr. Res.* **4**, 835–843.
Odler, I. & Dörr, H. 1979 *Cem. Concr. Res.* **9**, 277–284.
Powers, T. C. 1961 *J. Portland Cem. Ass. res. dev. Labs.* **3**, 47–56.
Previte, R. W. 1971 *Cem. Concr. Res.* **1**, 301–316.
Rendleman, J. A. 1966 *Adv. carbohyd. Chem.* **21**, 209–260.
Roy, D. M., Daimon, M. & Asaga, K. 1980 In *Proc. 7th Int. Congr. Chem. Cem., Paris*, vol. II, II.242–II.245. Paris: Éditions Septima.
Singh, N. B. & Ojha, P. N. 1980 In *Proc. 7th Int. Congr. Chem. Cem., Paris*, vol. II, II.100–II.105. Paris: Éditions Septima.
Skalny, J. P. & Young, J. F. 1980 In *Proc. 7th Int. Congr. Chem. Cem., Paris*, vol. I, II.1/3–II.1/35. Paris: Éditions Septima.
Slegers, P. A. & Rouxhet, P. G. 1977 *Cem. Concr. Res.* **7**, 31–39.
Stein, H. N. 1961 *J. appl. Chem.* **11**, 474–492.
Tadros, M. E., Skalny, J. F. & Kalyoncu, R. S. 1976 *J. Am. ceram. Soc.* **59**, 344–351.
Taylor, H. F. W. & Roy, D. M. 1980 In *Proc. 7th Int. Congr. Chem. Cem., Paris*, vol. I, II.2/1–II.2/12. Paris: Éditions Septima.
Tenoutasse, N. 1969 *Report Cembureau Colloquy, Paris*, 1–16.
Thomas, N. L. & Double, D. D. 1981 *Cem. Concr. Res.* **11**, 675–687.
Thomas, N. L. & Double, D. D. 1983 *Cem. Conr. Res.* **13**, 391–400.
Thomas, N. L., Jameson, D. A. & Double, D. D. 1981 *Cem. Concr. Res.* **11**, 143–153.
Williamson, R. B. 1972 *Prog. Mater. Sci.* **15**, 3–285.
Winslow, D. N. & Diamond, S. 1974 *J. Am. ceram. Soc.* **57**, 193–197.
Young, J. F. 1972 *Cem. Concr. Res.* **2**, 415–433.
Young, J. F., Tong, H. S. & Berger, R. L. 1977 *J. Am. ceram. Soc.* **60**, 193–200.

Discussion

P. C. Hewlett (*Cementation Research Ltd, Maple Cross, Denham Way, Rickmansworth, Herts, U.K.*). Would Dr Double care to comment on the reactions taking place under the *first* hydration peak and how they take part in a recombination reaction to form the C–S–H layer on the clinker grain?

D. D. Double. A substantial proportion of the heat in the first peak derives from the heat of wetting of the cement powder and the hydration of any free lime. These tend to mask contributions from the initial hydration of the cement phases so it is difficult to arrive at any firm conclusions regarding early hydration reactions.

K. S. W. Sing (*Department of Chemistry, Brunel University, Uxbridge, U.K.*). Our work has indicated that hydrated calcium ions are only partially desolvated when they are incorporated into freshly formed C–S–H gel. Water ligands are progressively removed during the localized development of gel structure. In our view the retention and removal of this ligand water plays a key role in this hydration process and the development of gel strength. Such an approach could help to explain some of the results presented by Dr Double.

D. D. Double. It is certainly true that desolvation effects are an important factor in the agglomeration of colloidal gels, and Profesor Sing's observations are very pertinent to considerations of cement chemistry.

J. F. Raffle (*University of Technology, Loughborough, U.K.*). Dr Double classified accelerators and retarders in terms of $\dot{Q}_{max}/t_{max}$. What happens to the width of this broad peak with different accelerators and retarders?

D. D. Double. Admixtures affect the heat evolution profiles in different ways. Generally, accelerators sharpen the second peak, increase the height of the peak maximum and bring it to shorter time intervals. Retarders depress and broaden the peak so that the integrated area under the heat curve, which gives a measure of the degree of reaction, is reduced.

Phil. Trans. R. Soc. Lond. A **310**, 67–78 (1983) [67]
Printed in Great Britain

Chemistry of colloidal silicates and cements

By R. H. Ottewill, F.R.S.
School of Chemistry, University of Bristol, Bristol BS8 1TS, U.K.

During the hydration of cement, colloidal particles are formed in the aqueous environment between the aggregate units. Their subsequent behaviour depends on the chemical nature of the particles and the additives present in the system. A review is given of the properties of colloidal particles likely to be present in the cement paste and the manner in which their properties are likely to influence the ultimate behaviour of the material.

1. Introduction

In the hydration of cement, reactions at interfaces must play a major role in determining the subsequent chemistry that occurs and ultimately the strength of the composite material formed. Moreover, the reactions that occur lead to the formation of colloidal particles and these also play an important part in the setting and ageing process (Taylor 1966; Double *et al.* 1978). Although the systems involved are complex, it is of considerable interest to examine this subject from a fundamental viewpoint to see whether this approach can provide guidance in the interpretation of existing phenomena and suggest how modifications might be made to produce materials with new properties. In this paper some of the salient points of the colloidal chemistry of inorganic materials are summarized and suggestions are made as to their applicability to cement hydration.

2. Behaviour of oxide surfaces

It is a general property of most oxide surfaces, if not all, that on contact with water, hydrolysis occurs leading to the formation of surface hydroxyl groups. The latter can then react with water to form either negatively or positively charged surface groups. This can be represented schematically in the following way:

$$\mathrm{>M{-}OH} \xrightarrow{H_2O} \mathrm{>M{-}OH_2^+} + \mathrm{OH^-}$$

$$\mathrm{>M{-}OH} \xrightarrow{H_2O} \mathrm{>M{-}O^-} + \mathrm{H_3^+O}$$

Since at the surface there are many charged groups per unit area it acquires an electrostatic potential, ψ_0, relative to the bulk solution (earth). To maintain the condition of electroneutrality, the charge on the surface has to be compensated by an equal and opposite charge in the solution phase. Thus, ions of opposite charge to the interfacial charge (counter-ions) are attracted towards the surface (positively adsorbed), while ions of opposite charge to the surface (co-ions) are repelled (negatively adsorbed). The distribution of ions in this region close to the

5-2

surface, the so-called diffuse electrical double layer, is given by the Boltzmann equation (Verwey & Overbeek 1948). On the basis of this simple model, for a planar interface with $\psi_0 < 25$ mV, the potential in the solution phase at a distance x from the surface is given by

$$\psi_x = \psi_0 \exp(-\kappa x).$$

The parameter κ is directly related to the electrolyte concentration in the bulk phase by the relation

$$\kappa^2 = 2n_0 v^2 e^2/\epsilon_r \epsilon_0 kT,$$

with n_0 = the number of ions of each kind per unit volume of the bulk phase, v = the magnitude of the valency of the ions, assuming a symmetrical electrolyte, ϵ_r = the relative permittivity of the solution phase and ϵ_0 that of free space; k = the Boltzmann constant and T = absolute temperature.

The quantity, κ, which has dimensions of reciprocal length, is important in that it moderates the fall off of potential with distance from the surface, i.e. when $x = 1/\kappa$, then $\psi_x = \psi_0/\mathrm{e}$. Hence at low electrolyte concentrations, for example 10^{-5} mol dm^{-3}, 1:1 electrolyte, $1/\kappa \approx 100$ nm and at 0.1 mol dm^{-3}, $1/\kappa \approx 1$ nm. The effect of the electric field from the particle extends to a distance of the order of $2/\kappa$, so that in 10^{-5} mol dm^{-3} electrolyte the fields of two particles overlap when their surfaces are separated by a distance of the order of 400 nm, a long range interaction; whereas in 0.1 mol dm^{-3} electrolyte the range of the interaction is two orders of magnitude smaller.

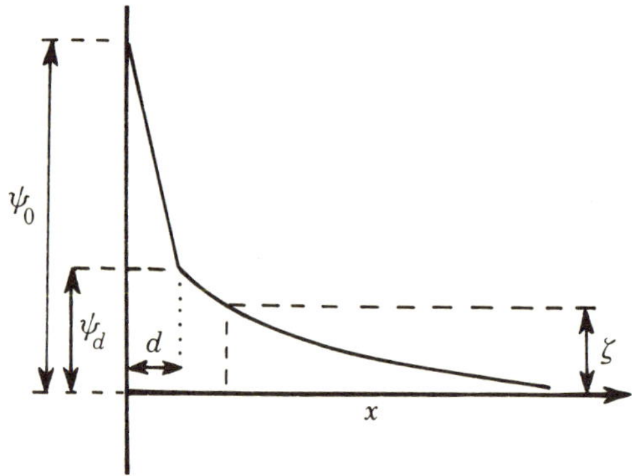

FIGURE 1. Schematic illustration of the electrical double layer.

A more sophisticated model of the electrical double layer includes the possibility of some counter-ions adsorbing in a layer adjacent to the charged surface and at a distance d from it, to form the so-called Stern layer (Stern 1924; Verwey & Overbeek 1948). The potential distribution on this basis is given in figure 1. There is now a linear drop between the surface and the inner layer followed by an exponential decrease with distance into the solution phase. The potential ψ_d is not easy to determine, but a quantity closely approaching it in magnitude, the zeta-potential, ζ, can be measured by electrokinetic experiments (Kruyt 1952). One of the simplest methods is to determine the electrophoretic mobility, u, by observing the movement of a particle, microscopically or ultramicroscopically, in an electric field (Hunter 1981). Then u is related to ζ by the expression (Wiersema *et al.* 1966)

$$u = (\epsilon_r \epsilon_0 \zeta/\eta) f(\kappa a, \zeta),$$

where η = the viscosity of the medium and $f(\kappa a, \zeta)$ is a numerical factor, which can be obtained from the literature (Ottewill & Shaw 1972; O'Brien & White 1978).

Some results obtained from silica and alumina particles as a function of pH in 10^{-3} mol dm^{-3} sodium chloride solution are shown in figure 2 (Ottewill & Holloway 1975). These show clearly that the charge at the surface of a silica particle is negative over a wide pH range and becomes increasingly negative as the pH is increased, i.e. it behaves as an acidic oxide. Alumina particles, however, are positively charged at pH values below *ca.* 9.0, and negative above this; alumina behaves as a basic oxide.

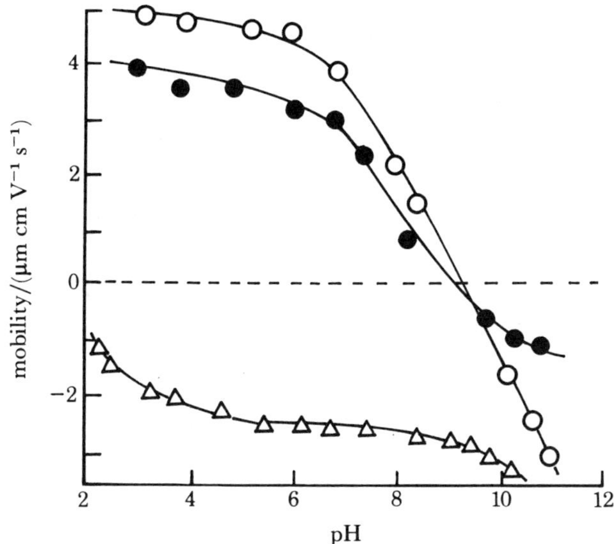

FIGURE 2. Mobility against pH. Key: △, silica in 10^{-3} mol dm^{-3} sodium chloride solution; ○, alumina in 10^{-3} mol dm^{-3} sodium chloride solution; ●, alumina in 10^{-1} mol dm^{-3} sodium chloride solution.

Figure 2 also shows the effect of electrolyte concentration on the electrophoretic mobility of alumina. Increasing the concentration of sodium chloride decreases the electrophoretic mobility, and hence the magnitude of ζ, by compression of the electrical double layer.

3. Equilibria between surface and solution

In examining the properties of a surface, particularly oxides as a function of pH, it is important to consider also the solution species that are likely to be in equilibrium with the surface. For many systems, stability constants are available (Martell & Sillen 1964) and hence solubility diagrams can be constructed. Examples of these are shown in figures 3 and 4 for three of the possible components of cement paste, silica as quartz (Stumm & Morgan 1970, pp. 395, 396), alumina (Gayer *et al.* 1958) and calcium hydroxide (Stumm & Morgan 1970, pp. 168, 169). From these diagrams it is clear that silica in the pH range of interest in cement pastes, 12.5–12.7 (Birchall *et al.* 1978), is appreciably soluble and that its solubility increases rapidly with pH above pH 10.0. Several silicate species also exist in solution and in addition the possible presence of polymeric species cannot be excluded since the rate of attainment of equilibrium in these systems is often slow (Iler 1979). Alumina particles are soluble in both acid and alkaline environments and it should be noted that the minimum solubility occurs at a pH of *ca.* 7.5, quite close to the isoelectric point (see figure 2). In fact, a close correlation exists between the

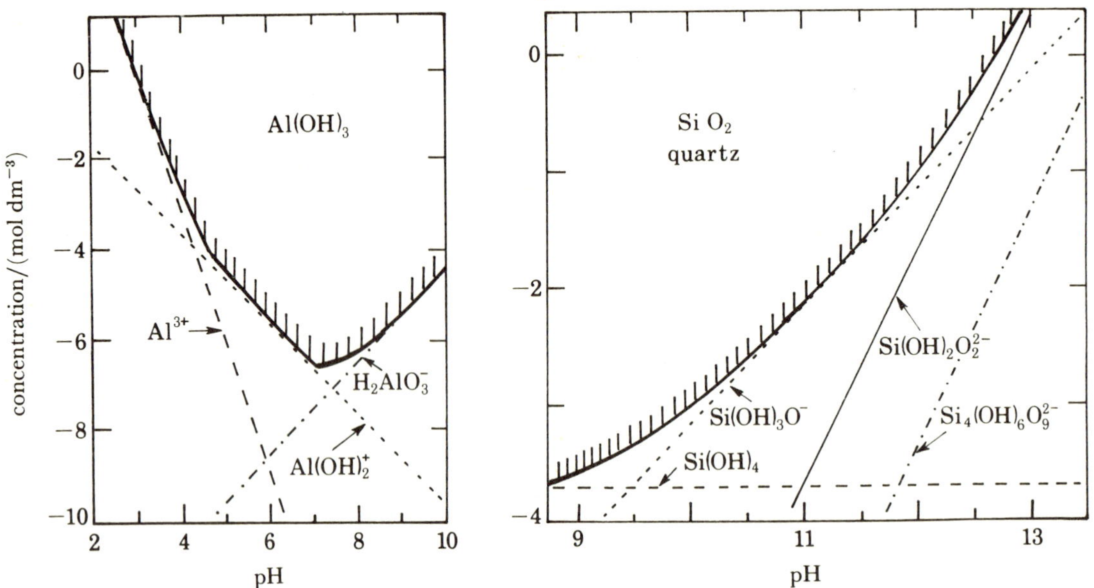

FIGURE 3. Solubility diagrams for alumina and quartz. (Logarithmic ordinate.)

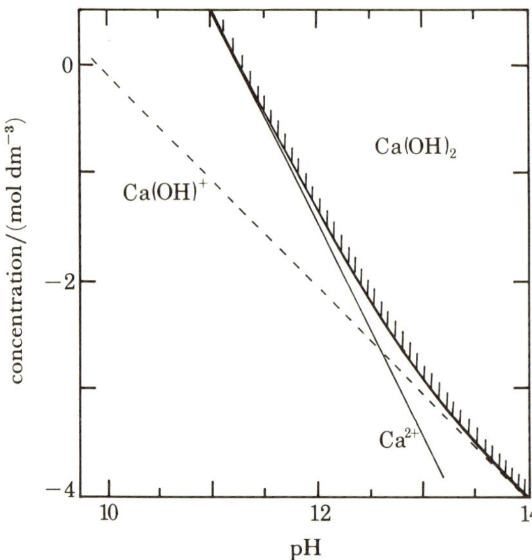

FIGURE 4. Solubility diagram for calcium hydroxide. (Logarithmic ordinate.)

solid solution equilibrium and the surface properties (Parks 1965; Stumm & Morgan 1970, pp. 474–493). The diagram for calcium hydroxide is an interesting one in that it shows, contrary to silica, a decreasing solubility as the pH increases and the presence of both Ca^{2+} and $Ca(OH)^+$ in the region of interest.

4. Ion exchange at surfaces

The simplest possible form of ion exchange is that between ions in the diffuse electrical double layer and the solution phase. If the ions have the same valency this occurs in direct proportion to their concentrations in the bulk phase. Exchange of a 2+ ion for a 1+ ion, however, is always in favour of the higher valent counter-ion because of the larger electrostatic factor, and hence is also dependent on the electrostatic potential (Kruyt 1952). Such exchange processes are non-specific in the chemical sense and are seldom found in practice.

To introduce specific effects, ion-exchange processes in the inner part of the double layer must be considered, and a chemical free energy of interaction with the surface introduced, in addition to the electrostatic term. The interaction of the adsorbed ion with the surface varies from simple electrostatic attraction to chemical bonding. In some cases the chemical reaction can be sufficiently extensive that metathesis occurs and the surface becomes coated with a layer of a new chemical compound.

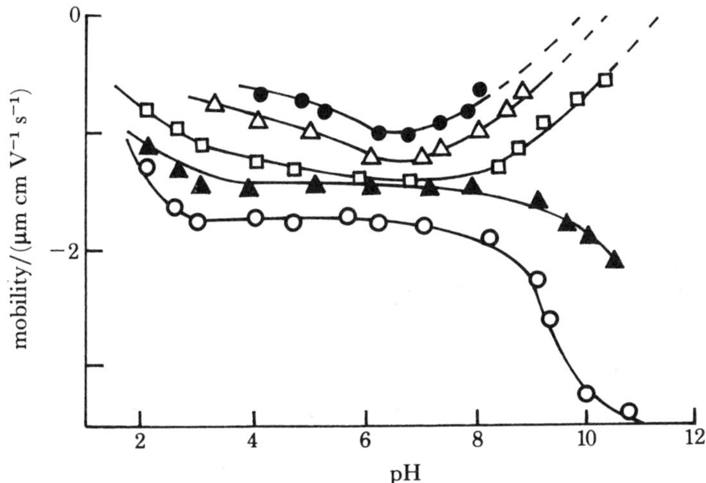

FIGURE 5. Mobility against pH for silica particles in various concentrations (mol dm^{-3}) of calcium chloride: ○, 10^{-4}; ▲, 10^{-3}; □, 10^{-2}; △, 5×10^{-2}; ●, 10^{-1}.

To demonstrate this effect we can examine the influence of calcium chloride on silica particles as a function of pH by using the electrokinetic results shown in figure 5. At low concentrations of calcium, 10^{-4} mol dm^{-3}, there is little effect and the behaviour is very close to that of silica (Ottewill & Holloway 1975). At 10^{-2} mol dm^{-3} calcium chloride, little effect other than double layer compression is observable below pH 7. Above this pH, however, the behaviour changes and the mobility starts to decrease rapidly. At about pH 10 (see figure 6) the system coagulates, i.e. it forms compact clumps. However, at a slightly higher pH flocculation occurs and a loose fluffy texture is observed. Increasing the calcium concentration still further enhances the effect above pH 7 and it appears that at 10^{-1} mol dm^{-3} extrapolation to pH 10 would give charge reversal of the surface, but coagulation made measurements in this region unreliable. Clearly, a reaction has occurred between calcium ions and the surface. The most likely reactions are:

$$\equiv\text{Si—O}^- + \text{Ca}^{2+} \rightarrow \equiv\text{Si—O}^- \ldots \text{Ca}^{2+}$$
$$\equiv\text{Si—O}^- + \text{Ca}^{2+} \rightarrow \equiv\text{Si—O Ca}^+$$
$$\equiv\text{Si—O}^- + \text{Ca(OH)}^+ \rightarrow \equiv\text{Si O Ca OH}.$$

Alternatively, reaction could also occur with remaining unionised hydroxyl groups

$$\equiv\text{Si—OH} + \text{Ca}^{2+} \rightarrow \equiv\text{Si O Ca}^+ + \text{H}^+$$
$$\equiv\text{Si—OH} + \text{Ca(OH)}^+ \rightarrow \equiv\text{Si O Ca(OH)} + \text{H}^+$$
$$\equiv\text{Si—OH} + \text{Ca(OH)}^+ \rightarrow \equiv\text{Si O Ca}^+ + \text{H}_2\text{O}$$

The fact that the net charge on the particles is reduced and that Ca^{2+} is the dominant ion in solution probably favours the first three equations although it is recognized that hydroxylated ions have a high affinity for surfaces (Matijević 1977) and could hydrogen-bond to surface hydroxyls. Although the electrokinetic technique does not in itself elucidate the actual reaction, it does provide direct evidence that a reaction has occurred. That some ion exchange at the surface can also occur, is demonstrated by the results given in figure 6. Addition of sodium chloride causes an increase in the magnitude of the negative mobility, which suggests the displacement of calcium ions by sodium ions.

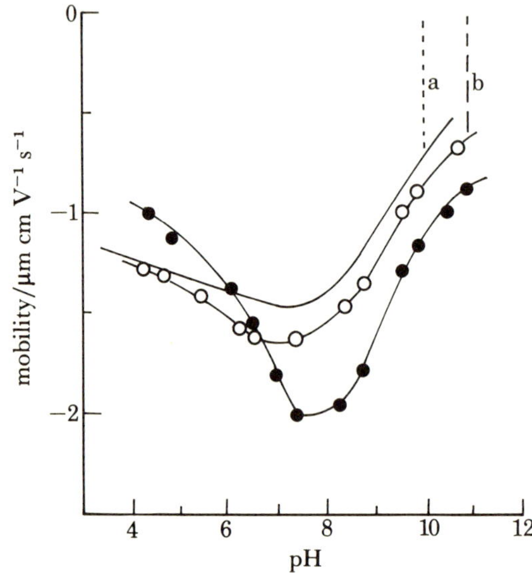

FIGURE 6. Mobility against pH for silica particles in the presence of 10^{-2} mol dm^{-3} calcium chloride: —, sodium chloride absent; ○, 10^{-2} mol dm^{-3} sodium chloride; ●, 10^{-1} mol dm^{-3} sodium chloride; a, coagulation, and b, flocculation.

5. THE STABILITY OF COLLOIDAL DISPERSIONS

The basis of current theories of colloid stability is to consider the interaction between two particles in terms of the electrostatic repulsion, V_r, and the van der Waals attraction, V_a, as a function of the distance of surface separation, h, as illustrated in figure 7. For two spherical particles of identical radius, a, V_r is given by the equation

$$V_r = 2\pi\epsilon_r\epsilon_0 a\psi_d^2 \ln\left[1+\exp\left(-\kappa h\right)\right],$$

and V_a as a first approximation by

$$V_a = -Aa/12h,$$

where A is a material constant for the system known as the Hamaker constant (Hamaker 1937; Mahanty & Ninham 1976). For practical purposes ψ_d is often replaced by ζ. A full discussion of these equations can be found elsewhere (Verwey & Overbeek 1948; Ottewill 1977, 1982; Smith 1969). For the present purposes we can assume that the total interaction energy, V_t, is given by

$$V_t = V_r + V_a + V_B,$$

where V_B is a very close range Born repulsion. From these equations we can examine the influence of ionic strength and diffuse layer potential on the form of the potential energy of interaction. Some schematic energy curves are given in figure 8. It is immediately clear from 8*a* that if salt is added to a system, at constant ψ_d, the maximum is depressed until at *ca.* 0.1 mol dm^{-3} it occurs at *ca.* $V_t = 0$, or in other words, since the particles no longer have to surmount a kinetic energy barrier, they can go into a deep energy minimum and come into close contact where they touch. Thus the onset of instability in colloidal dispersions can be correlated with the depression of the kinetic energy barrier to association. This can be achieved by addition of salt (compression of the electrical double layer), by reduction of the surface potential (specific adsorption), or by a combination of both effects.

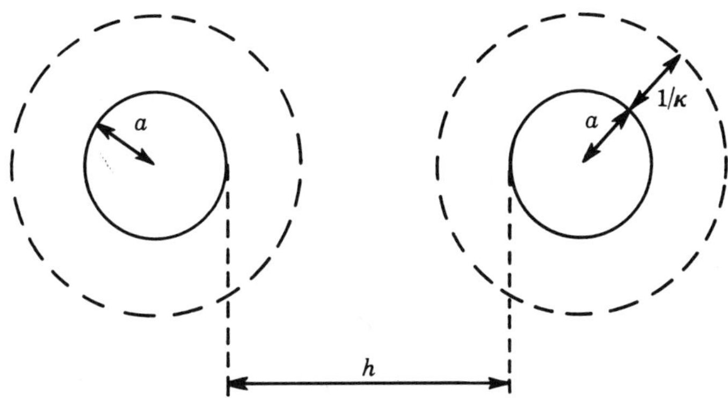

FIGURE 7. Interaction between two spherical particles.

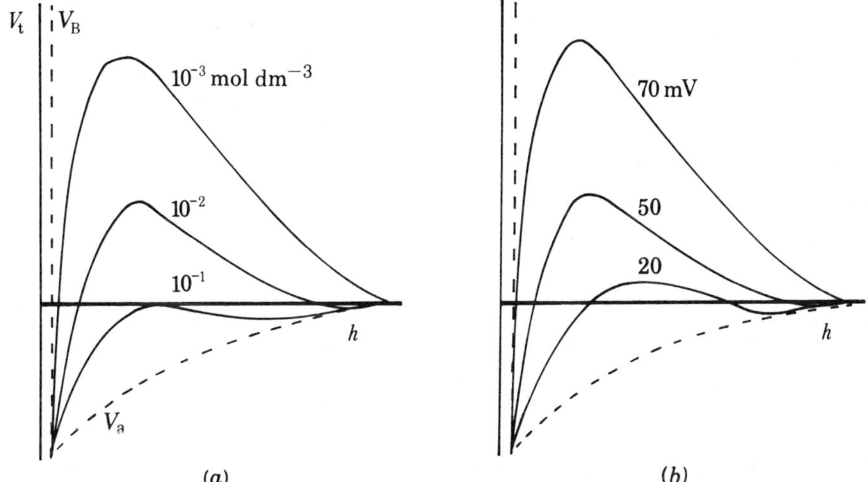

FIGURE 8. Schematic potential energy diagrams: (*a*) effect of salt concentration; (*b*) effect of surface potential.

The arguments advanced show quantitatively that the electrolyte concentration at which loss of stability occurs, the so-called critical coagulation concentration, is dependent inversely on a power of the valency ($1/v^n$), where $n = 2$ for coagulation at low potentials and $n = 6$ for coagulation at high potentials (Ottewill 1982). This provides a theoretical basis for the valency dependence in coagulation known as the Schulze–Hardy Rule (Kruyt 1952).

However, as we have discussed earlier, ion exchange can occur in the inner part of the electrical double layer and lead to specific effects of the ions at surfaces. Thus ions of the same valency are frequently found to form sequences which reflect their own properties, e.g. extent of hydration and polarizability, and their interaction with the surface. A good example is found with the alkali metal ions where the concentration required to produce coagulation is frequently in the order $Li^+ > Na^+ > K^+ > Rb^+ > Cs^+$, the lyotropic series (Kruyt 1952).

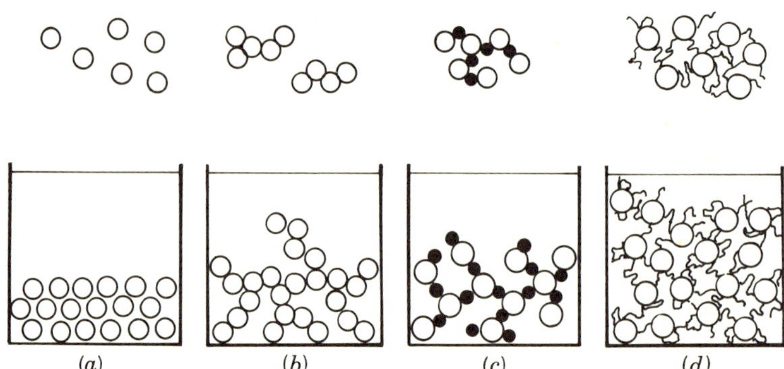

FIGURE 9. States of colloidal dispersions (upper sketch) and settled volumes (lower sketch): (*a*) stable dispersion, (*b*) coagulation, (*c*) heterocoagulation, (*d*) flocculation.

Another important concept that arises from these considerations is the way in which particles pack in concentrated dispersions. The repulsive forces are dominant for stable systems; these keep the particles apart, maintain a fluid layer between them, and hence allow the particles to assemble, say in sedimentation, into a close packed array with a low sedimentation volume. Under coagulation conditions the particles contact at the point where they kinetically impact. Furthermore, the liquid film is essentially lost, leading to contacts with high friction, and hence the particles pack to form a disordered array with a high void space. These effects are illustrated in figure 9.

Two other considerations can be added. First, if positively charged particles are added to negatively charged particles, the phenomenon of heterocoagulation is observed (figure 9). This is also an effect that can sometimes be obtained with particles of the same sign of charge but with different potentials and different sizes. Second, if polymeric species are present in a system, either by deliberate addition or by formation *in situ*, these can link the particles by a 'bridging' mechanism at low concentrations or 'protect' the particles by extensive adsorption at higher concentrations (Napper 1982). Addition of polymers frequently leads to flocculation, a state in which the primary particles remain separated by the liquid phase but are linked by the polymer. The consequence is the formation of units of high volume requirements, i.e. a high sediment volume, which visually often have a woolly or fluffy appearance.

The illustrations given in figure 9 are an attempt to illustrate schematically the different states that can be realized in practice. It should also be emphasized that combinations of these can occur.

Figure 10 shows the change in sedimentation volume of silica particles in the presence of 5×10^{-2} mol dm^{-3} calcium chloride as the pH is increased. Taking the sedimentation volume at pH 8 as unity, it can be seen that above pH 9.5 an increase occurs, which rises rapidly above pH 10.6 (Holloway 1974). These results can be correlated with figure 6, which indicates that as pH 9.5–10.0 is approached the electrophoretic mobility approaches zero and hence coagulation would be expected. Above pH 10.0, reversal of charge by the adsorption of calcium would be expected, leading to the formation of positively charged particles. Moreover, as the pH rises dissolution of the silica must occur (figure 3) to give silicate species in the solution phase, probably polymeric, which lead to flocculation.

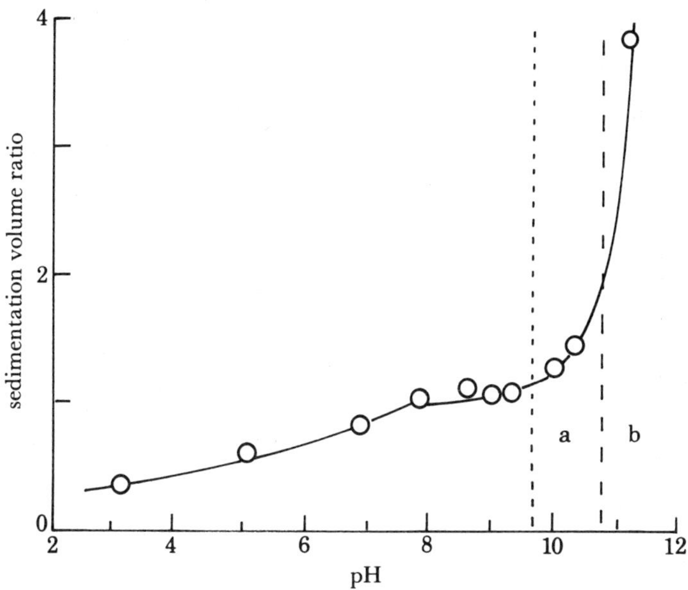

FIGURE 10. Sedimentation volume ratio against pH for silica in 5×10^{-2} mol dm^{-3} calcium chloride solution; a, coagulation, and b, flocculation.

6. Formation of new particles

If two soluble species A and B are mixed to give a reaction

$$A + B \rightarrow C,$$

where C is an insoluble compound, then initially the concentration of C increases with time, $C(t)$, as shown in figure 11. The concentration exceeds the solubility limit but new particles are not formed at this point. Nucleation only occurs once the critical supersaturation value has been reached and the rate of nucleation, R_n, becomes finite. Once nucleation occurs the solution concentration falls. In fact, it will fall back, after a short time interval Δt, below the critical supersaturation so that no further nucleation occurs and growth on the existing particles occurs by diffusion of the solute species on to the existing nuclei. If the formation of nuclei is over a small time span, Δt, then monodisperse particles can be formed. More frequently $C(t)$ rises until $R_n \rightarrow \infty$ and nucleation lasts over a long time period thus giving very heterodisperse dispersions.

The fate of the particles formed depends on the environment in which they find themselves, for example, just to take a few cases,

(i) if there is a low salt concentration and the particles have a high ψ_d then a stable colloidal dispersion would be formed;

(ii) if there is a high salt concentration and the particles have a low ψ_d then the particles would coagulate as they formed;

(iii) if the particles are formed in the presence of a polymeric species, or are generated simultaneously with a polymer species, then a flocculated dispersion could well result;

(iv) if nucleation occurs in the presence of existing particles it could occur preferentially at the surface of those particles (*surface* nucleation), leading to overcoating or the growth of new particulate units on the surface.

In the formation of highly hydroxylated materials it is frequently found that the resulting dispersion, whether stable or unstable, is composed of amorphous units. These can be highly hydrated and contain ions. With time, the units reassemble to form new crystalline particles. Ostwald ripening, the preferential dissolution of small particles with growth of the larger particles, also plays a role in this process. Several changes can occur, frequently slowly, until the most stable thermodynamic species is formed.

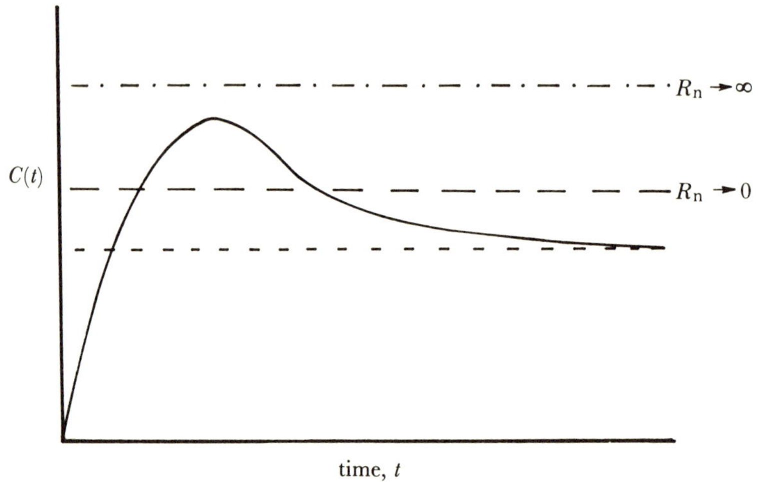

FIGURE 11. Concentration against time for a precipitating system: – – – –, solubility; — — —, critical super-saturation curve; · — · —, critical limiting supersaturation.

7. Comments on cement hydration

The background considerations given in the previous sections can now be applied to speculate about the processes that might occur during the hydration of cement. As soon as the cement comes into contact with water it is apparent that the pH rises rapidly to *ca.* 12.5 (Birchall *et al.* 1978). At this pH both silica and calcium hydroxide are soluble. Hence, if the particles are separated we can visualize diffusion processes (figure 12) in which soluble silica species diffuse outwards from the silica particles and Ca^{2+} and $Ca(OH)^{+}$ diffuse from the calcium hydroxide. These ions must meet in a zone in which nucleation can occur to form a colloidal dispersion. Whether this is close to the silica or the calcium hydroxide will depend on the rate of dissolution of each and this rate will also be dependent on factors such as, the chemical nature of the particles, particle size, surface area, temperature, porosity etc.

The silica and calcium hydroxide will be present in the same aggregate unit together with tricalcium silicate, dicalcium silicate etc., for cement, so that the situation is more complex. It has already been demonstrated that calcium has a high affinity for silica surfaces at high pH. The fact that the calcium content of the solution rises rapidly with time, whereas little silica is found in solution (Thomas & Double 1981), might well suggest that the nucleation zone for colloidal particle formation is near to the silica grain surface and that in the high electrolyte concentration the particles thus formed would immediately flocculate to give a gel-like layer at, or near to, the surface. With continuous diffusion this region would continue to grow and expand away from the surface. Also, while the dissolution continued under attack by OH^- ions, there would accumulate an ion imbalance between this interphase and the solution, leading to osmotic gradients, as suggested by the osmotic membrane hypothesis (Birchall *et al.* 1978; Double *et al.* 1978; Birchall 1980). As a function of time, crystallization of this amorphous material might reasonably be expected to occur and to lead to a contraction and the creation of void space. The role of carbon dioxide must also be considered since this could also lead to the formation of calcium carbonate particles.

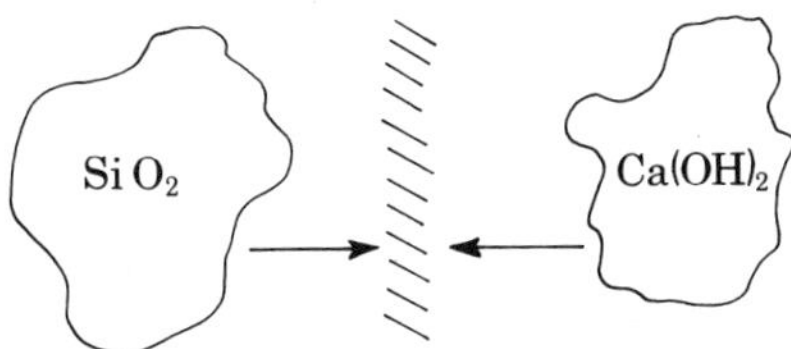

FIGURE 12. Schematic diagram showing formation of a precipitation zone.

It seems likely that since silica is the more slowly dissolving species, it is this factor that is the rate controlling process. It is also possible that the attack on the silica by OH^- ions occurs at or near the surface and for this reason material is pushed outwards in the form of needle-like growths. Moreover, since we are suggesting a particle association process, probably with bridging by polysilicic species, the gel matrix so formed is likely to be amorphous and to have a variable composition.

In the light of this suggested mechanism some indication should be possible as to the likely effect of additives. In terms of electrolyte addition it would be expected that valency effects would be important in controlling coagulation and that these should give a Schulze–Hardy dependence but with some ions showing a pronounced specificity. Materials such as ionic surface active agents, provided they are stable in the alkaline region, would be expected to have a pronounced effect on the nucleation, leading to the formation of smaller particles. In addition there would be adsorption on the larger aggregates, which would help to maintain a fluid layer between the macroscopic surfaces and aid flow and particle packing.

The effect of polymer addition would depend on a number of factors, such as, molecular mass, charge type, degree of branching etc. In the nucleation phase the effect would be uncertain in that either flocculation or stabilization could occur according to conditions. With the larger particles, i.e. large in comparison with the size of the polymer, adsorption of the polymer at the inorganic interface would be expected to give a layer with a thickness of the order of 10 nm or greater (Eirich 1977; Bangham *et al.* 1979) with a high degree of hydration. Macromolecules such as hydrolysed polyvinylacetate or carboxymethylcellulose, which have been shown to be effective as steric stabilizers for aqueous dispersions, might be expected to aid flow by preventing the inorganic solids of high surface energy from coming into intimate contact. These materials, as shown by Birchall *et al.* (1983), are likely to produce much better particle packing thus reducing void volume and ultimate porosity.

I wish to express my sincere thanks to Professor J. D. Birchall, F.R.S., for introducing me to this topic and for a number of stimulating discussions on the subject.

References

Bangham, O. B., Maternaghan, T. J. & Ottewill, R. H. 1979 *Photogr. Sci. & Engng* **23**, 45–52.
Birchall, J. D. 1980 In *Symposium: Advances in Cement Matrix Composites, Boston*, U.S.A.
Birchall, J. D., Howard, A. J. & Bailey, J. E. 1978 *Proc. R. Soc. Lond.* A **360**, 445–453.
Double, D. D., Hellawell, A. & Perry, S. J. 1978 *Proc. R. Soc. Lond.* A **359**, 435–451.
Eirich, F. R. 1977 In *Colloid and interface science* (ed. M. Kerker) vol. I, pp. 447–460. New York: Academic Press.
Gayer, K. H., Thompson, L. C. & Zajicek, D. T. 1958 *Can. J. Chem.* **36**, 1268–1271.
Hamaker, H. C. 1937 *Physica* **4**, 1058–1072.
Holloway, L. R. 1974 Ph.D. thesis, University of Bristol.
Hunter, R. J. 1981 *Zeta potential in colloid Science*. London: Academic Press.
Iler, R. K. 1979 *The chemistry of silica*. New York: John Wiley and Sons.
Kruyt, H. R. 1952 *Colloid Science*, vol. I. *Irreversible Systems*. Amsterdam: Elsevier.
Mahanty, J. & Ninham, B. W. 1976 *Dispersion forces*. London: Academic Press.
Martell, A. E. & Sillén, L. G. 1964 *Stability constants of metal–ion complexes*. London: The Chemical Society.
Matijević, E. 1977 In *Colloid and interface science* (ed. M. Kerker) vol. I, pp. 397–412. New York: Academic Press.
Napper, D. H. 1982 In *Colloidal dispersions* (ed. J. W. Goodwin) pp. 99–128. London: Royal Society of Chemistry.
O'Brien, R. W. & White, L. R. 1978 *J. Chem. Soc. Faraday Trans.* II **74**, 1607–1626.
Ottewill, R. H. 1977 *J. Colloid Interface Sci.* **58**, 357–373.
Ottewill, R. H. 1982 In *Emulsion polymerization* (ed. I. Piirma) pp. 1–49. New York: Academic Press.
Ottewill, R. H. & Holloway, L. R. 1975 In *The nature of sea water* (ed. E. D. Goldberg) pp. 599–621. Berlin: Dahlem Konferenzen.
Ottewill, R. H. & Shaw, J. N. 1972 *Electroanal. Chem. interfacial Electrochem.* **37**, 133–142.
Parks, G. A. 1965 *Chem. Rev.* **65**, 177–198.
Smith, A. L. 1969 In *Dispersion of powders in liquids* (ed. G. D. Parfitt) pp. 39–80. Amsterdam: Elsevier.
Stern, O. 1924 *Z. Elektrochem.* **30**, 508.
Stumm, W. & Morgan, J. J. 1970 *Aquatic chemistry*. New York: Wiley-Interscience.
Taylor, H. F. W. 1966 *R. Inst. Chem. Lect. Ser.* no. 2, pp. 1–27.
Thomas, N. L. & Double, D. D. 1981 *Cem. Concr. Res.* **11**, 675–687.
Verwey, E. J. W. & Overbeek, J. Th. G. 1948 *Theory of stability of lyophobic colloids*. Amsterdam: Elsevier.
Wiersema, P. H., Loeb, A. & Overbeek, J. Th. G. 1966 *J. Colloid Interface Sci.* **22**, 78–99.

Discussion

K. S. W. Sing (*Brunel University, Uxbridge, U.K.*). Complex ageing mechanisms are involved in the hydration of Portland cement and this gives rise to changes in the microstructure over long periods. The classical process of Ostwald ripening appears to operate in the growth of the calcium hydroxide (Portlandite) crystals, but cannot explain the changes in C–S–H gel structure. The observed *increase* in the area of the solid–liquid interface appears to be due to the development of short-range order, which accompanies the removal of the water ligands. It seems to me important to attempt to distinguish between crystal growth and the ageing of poorly ordered gels such as C–S–H.

R. H. Ottewill. Probably the most important aspect of cement ageing, as in many inorganic materials that precipitate as gels, is the slow transformation of amorphous materials into crystalline products.

J. Bensted (*Blue Circle Industries p.l.c., Research Division, Greenhithe, U.K.*). Can Professor Ottewill comment on the effect of temperature on the coagulation of colloidal silicates? Has he any experience of the colloid chemistry of silicon having 6 co-ordination by hydroxyl?

R. H. Ottewill. On the coagulation process alone temperature effects are usually small because they scale with kT, and on the absolute scale of temperature there is only a small change between say ambient temperature and 50° C. On the other hand, processes such as the transformation of amorphous gels to crystalline products, which involve reactions leading to the elimination of water molecules for the product initially formed, can be temperature sensitive.

Phil. Trans. R. Soc. Lond. A **310**, 79–83 (1983)
Printed in Great Britain

T.e.m. studies of cement clinker compounds, their hydration and strong cement pastes

By G. W. Groves
Department of Metallurgy and Science of Materials, University of Oxford, Parks Road, Oxford, OX1 3PH, England, U.K.

[Plates 1 and 2]

Ion-beam thinned specimens of cement-forming compounds and cement pastes have been examined by transmission electron microscopy (t.e.m.). The triclinic lattice of pure tricalcium silicate has been confirmed and twins corresponding to domains of the superlattice have been observed. Thinned specimens of tricalcium aluminate show a copious amorphous reaction product at the surface after dipping into alkaline calcium sulphate solution, confirming the mechanism of retardation by blocking access to the surface. Observations of tricalcium silicate after it has been dipped into solution in contact with hydrating tricalcium silicate both with and without calcium chloride suggest that calcium chloride additive may accelerate hydration by improving access to the surface. Cement pastes of high strength ('macro-defect-free' pastes) show a similar fine-scale microstructure to cement pastes of low water content made by conventional techniques.

1. Introduction

A number of observations, by transmission electron microscopy (t.e.m.), of ion-beam thinned specimens of cement clinkers, cement forming compounds and cement pastes have recently been reported. Jennings *et al.* (1981) and Dalgleish *et al.* (1980) studied pastes of tricalcium silicate (C_3S) of various ages, including quite young pastes reinforced by resin for the purpose of thinning, and identified different forms of hydration product. Groves (1981 *a*) made a study of the microstructures of the major phases of ordinary Portland cement clinker and reported on the nature and origin of twinning in β-dicalcium silicate (βC_2S) (Groves 1982). Hudson & Groves (1981) examined the alite (impure C_3S) phase in ordinary Portland cement clinkers and showed that the majority of alite had the monoclinic structure determined by Jeffery (1952), and was twinned.

In the present study a variety of new observations are reported. Pure C_3S and tricalcium aluminate (C_3A) have now been examined and early stages of their hydration in water and other solutions have been studied by immersing a thinned specimen for a fixed period and then observing it by t.e.m. after withdrawal from solution. Cement pastes of low water/cement ratio (w/c) are particularly suitable for ion-beam thinning because of their relatively high strength. The microstructure of the recently developed 'macro-defect-free' (MDF) cement pastes has been compared with that of other ordinary pastes of low water content.

2. Results and discussion

(*a*) *The lattice and microstructure of pure tricalcium silicate* (C_3S)

An examination was made of pure C_3S in the form of flux-grown crystals kindly supplied by the National Bureau of Standards, Washington D.C., and also in the form of sintered specimens prepared in our own laboratory. Specimens were ion-beam thinned by standard techniques. Both materials showed the same electron diffraction patterns, which were quite distinct from patterns obtained from alites (impure C_3S) in cement clinker (Hudson & Groves, 1981). A prominent pattern is shown in figure 1, plate 1. From this and two other patterns from different orientations the reciprocal lattice was deduced and inverted to form the true direct lattice. This lattice was in complete agreement with that reported by Golovastikov *et al.* (1975), which was based on X-ray rotation photographs of single crystals. The lattice is triclinic, $a = 1.167$ nm, $b = 1.424$ nm, $c = 1.372$ nm, $\alpha = 105° 30'$, $\beta = 94° 20'$, $\gamma = 90°$. Lattices suggested in earlier work on the basis of X-ray powder photography (as listed, for example, in Taylor 1964) were found to be inapplicable, except with regard to the pseudolattice. The same pseudolattice, nearly trigonal, $c = 2.5$ nm, $a = 0.7$ nm, occurs for all forms of C_3S, including alites. The strong pseudolattice reflexions can be easily picked out in figure 1, and two of them are indexed on the plate. The remaining reflexions, which may be termed superlattice reflexions are relatively weak although they approach the pseudolattice reflexions in intensity as the specimen thickness increases. The ease with which even very weak superlattice reflexions are observed in t.e.m. is a great advantage in identifying the true lattice. Pure C_3S contains reflexion twins on the mirror planes of the pseudolattice, $\{2\bar{1}\bar{1}0\}$, which may be considered as domains in an originally trigonal structure, within which different but equivalent orientations are chosen for the superlattice planes. An example of this type of twinning is shown in figure 2. The weakness of the contrast between the twins is presumably a result of their giving the same strong pseudolattice reflexions, as can be seen in the diffraction patterns shown in the plate. These twins are exactly similar in character to those reported in alite, which has a quite different superlattice but the same pseudolattice as pure C_3S (Hudson & Groves 1981).

(*b*) *Early hydration of tricalcium aluminate* (C_3A)

The hydration products formed at the surface of thinned pure C_3A specimens, after dipping the specimen for a short time in water and in a solution co-saturated with calcium sulphate and calcium hydroxide, were compared. After immersion in water for 5 min the specimen, when removed from the water and allowed to dry, appeared to be remarkably clean with perhaps a trace of product at the edge. No diffuse diffraction ring or reflexions other than those of pure C_3A could be detected in diffraction patterns taken from regions at the specimen edge. The alkaline calcium sulphate solution, which was meant to simulate approximately the environment of C_3A in a Portland cement (retarded with calcium sulphate) shortly after the addition of water, produced a copious reaction product, after a similar 5 min immersion period. The product obscured the edge of the specimen completely (figure 3). In agreement with the findings of Bailey & Chescoe (1980), the reaction product appears from diffraction patterns to be completely amorphous, in spite of its somewhat fibrous morphology. The observations support the view, which is now becoming widely accepted, that the calcium sulphate acts as a retarder for the hydration of C_3A by very quickly (within *ca.* 5 min) smothering its surface with reaction products.

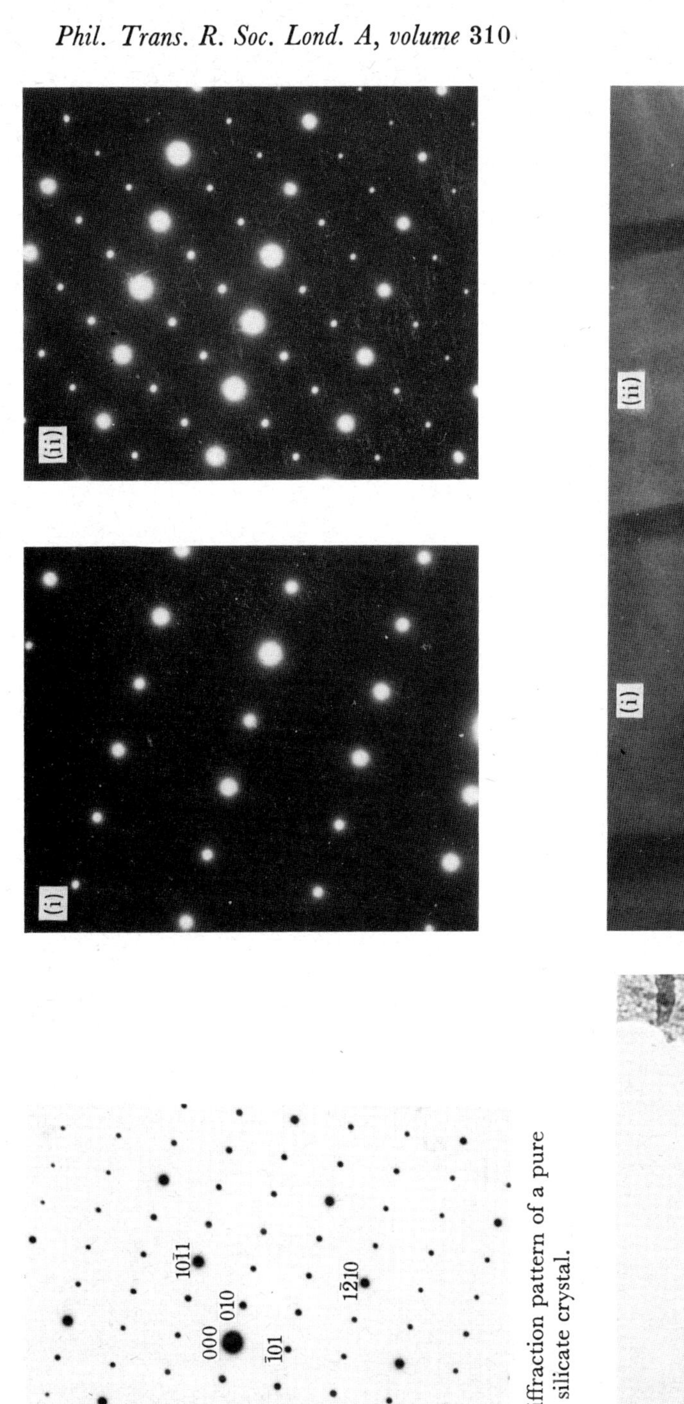

FIGURE 1. Electron diffraction pattern of a pure tricalcium silicate crystal.

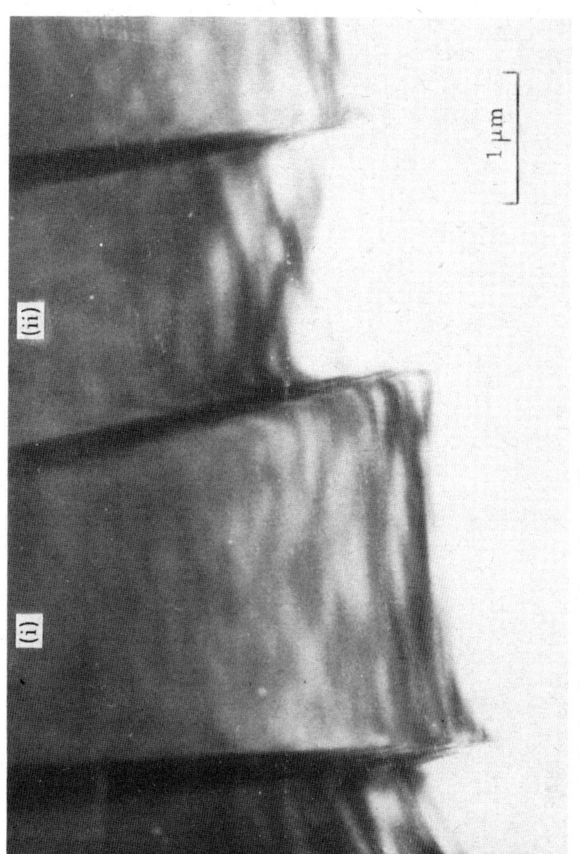

FIGURE 2. Twins in pure tricalcium silicate, with electron diffraction patterns from the regions (i) and (ii).

FIGURE 3. Amorphous surface coating formed on tricalcium aluminate immersed in alkaline calcium sulphate solution for 5 min.

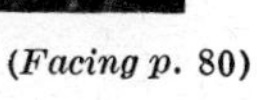

(*Facing p.* 80)

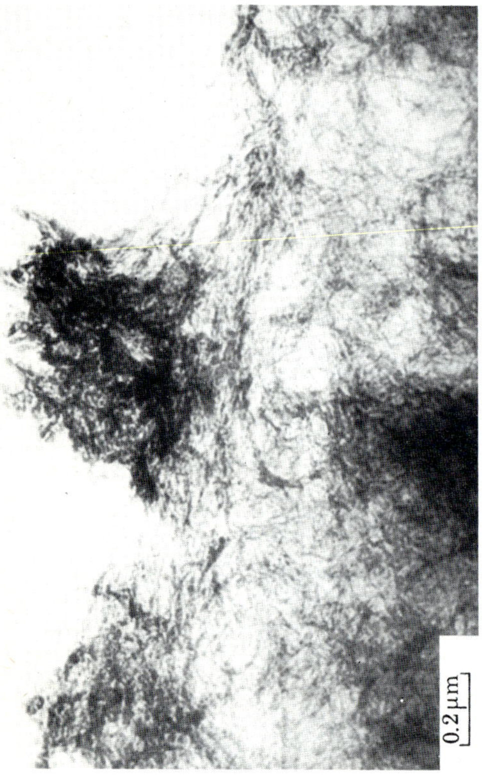

0.2 μm

FIGURE 4. Edge of a tricalcium silicate specimen after immersion in water for 3 h.

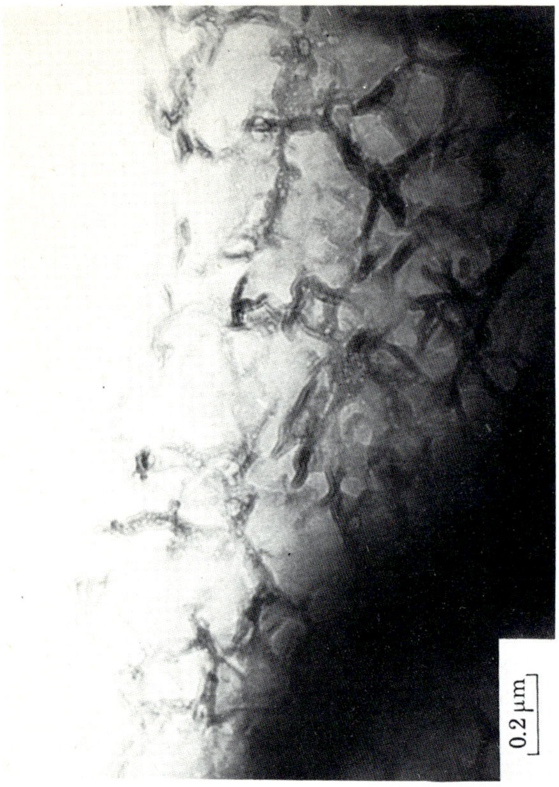

0.2 μm

FIGURE 5. Edge of a tricalcium silicate specimen after immersion in calcium chloride solution for 3 h.

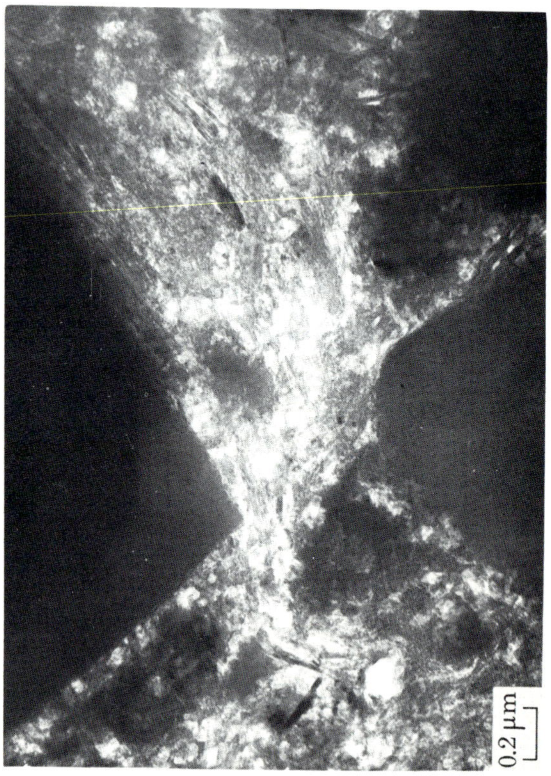

0.2 μm

FIGURE 6. Macro-defect-free cement paste microstructure. The large dark regions are residual cement clinker.

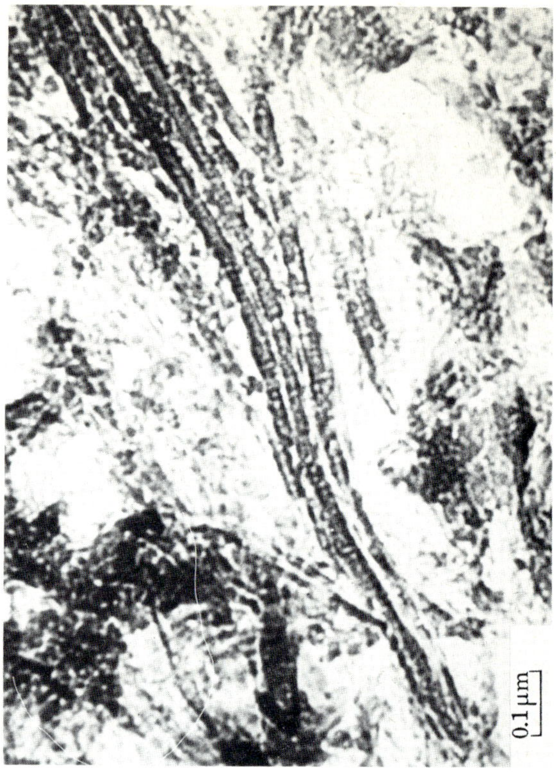

0.1 μm

FIGURE 7. Lamellae of calcium hydroxide embedded in hydrate gel in a macro-defect-free cement paste.

(c) *Early hydration of pure* C_3S

Thinned specimens were examined after immersions of 3 h and 6 h in the supernatant liquid of a 2:1 w/c ratio paste made with pure C_3S powder, the paste being mixed immediately before the immersion. After immersion the specimens were rinsed in water, drained and allowed to dry. For comparison, specimens were immersed for the same times in the liquid of a similar paste made with a 2% solution of the accelerator calcium chloride.

After 3 h in water a quite copious reaction product could be seen (figure 4, plate 2). Diffraction patterns normally showed both sharp reflexions and a diffuse ring, indicating amorphous material lying on crystalline C_3S, but an aperture selecting only the regions within *ca.* 0.3 μm of the edge produced only the diffuse ring pattern indicating an amorphous coating standing out from the edge. After 6 h immersion in water a very prominent amorphous zone *ca.* 1 μm in width was visible at the edge of the specimen. An immersion of 3 h in the accelerated cement solution left some amorphous material on the specimen surface (figure 5) but it appeared to be less copious than it was after immersion in the non-accelerated solution. After 6 h the microporous regions visible in figure 5 had spread to cover the surface but it was not possible to obtain a diffraction pattern from the extreme edge showing only a diffuse ring. Crystalline reflexions were always visible in addition to the diffuse ring. This indicated that there was not a coating standing out from the edge.

Although many further experiments need to be performed to establish the consistency of these observations, the preliminary results lead to the tentative suggestion that the role of calcium chloride in accelerating the hydration of C_3S may be approximately the reverse of the role of calcium sulphate in retarding the hydration of C_3A, i.e. that it acts by increasing the accessibility of the surface to hydration. Exactly how this comes about is not yet clear.

(d) *Microstructure of strong pastes*

The detailed microstructure of pastes made with the addition of a water-soluble polymer and processed to give MDF cement appears to be indistinguishable from that of other pastes made with a low water content, but not using water-soluble polymer. All the pastes examined have been relatively young – typically about one month old – and contain large amounts of unhydrated clinker, which unfortunately thins more slowly under the ion beam than the hydration products. This leads to a specimen of very irregular thickness, and it is a great advantage to examine such specimens under a 1 MV electron microscope, with its ability to penetrate thicker regions of the specimen.

A typical region of a MDF paste is shown in figure 6. This paste was mixed with water-soluble polymer and passed between rollers, following the type of process described in I.C.I. patents (Birchall *et al.* 1983). It was then hydrated for 4 weeks at 50 °C and because of the elevated temperature of hydration is probably more hydrated than is usual for MDF paste although there was no striking difference in microstructure between this paste and one hydrated at room temperature. A prominent feature of the microstructure is the presence of a dense homogeneous region of amorphous gel round the edge of clinker grains. Such regions have been seen in other pastes, for example C_3S pastes made in the normal way (Jennings *et al.* 1981), so they are clearly not a special feature of MDF pastes. Outside these regions the hydration product is an intimate mixture of crystalline and amorphous material. Diffraction from a region of 1 μm diameter in this outer hydration product nearly always shows a diffuse ring together with some crystalline reflexions, and regions showing diffraction contrast within the outer hydration product can usually be detected by tilting the specimen. One source of crystalline material is calcium hydroxide, which is present in a microcrystalline form (Groves 1981*b*). This consists of groups of very thin lamellar crystals, the plane of the lamellae being parallel to the basal plane (0001) of calcium hydroxide. The lamellae tend to lie approximately parallel to one another but are not otherwise oriented with respect to one another. A very small group of calcium hydroxide microcrystals is shown in figure 7, taken from the same cement as figure 6. Again, microcrystalline calcium hydroxide is not a special feature of MDF pastes but is detectable in other pastes of low water content.

3. Conclusions

(*a*) Electron diffraction patterns confirm that the lattice of pure C_3S is that found by Golovastikov *et al.* (1975). Twins are observed that can be regarded as domains of different superlattice orientation within the trigonal pseudolattice.

(*b*) Observations of thinned specimens after they had been dipped into appropriate solutions confirm that calcium sulphate retards the hydration of C_3A by forming a copious amorphous reaction product at its surface, and suggest that calcium chloride may accelerate the hydration of C_3S by reducing the amount of reaction product adhering to the surface.

(*c*) Observations of MDF cement pastes show that their fine-scale microstructure is similar to that of other cement pastes of low water content made in a conventional way. This is consistent with the view that the relatively high strength of MDF pastes is due to the absence of the larger pores present in conventional pastes (Birchall *et al.* 1983).

References

Bailey, J. E. & Chescoe, D. 1980 In *Proc. 7th Int. Congr. Chem. Cem., Paris*, vol. 4, p. 595.
Birchall, J. D., Howard, A. J. & Kendall, K. 1983 *Metallurgist. Mater. Technol.* **15**, 35.
Dalgleish, B. J., Pratt, P. L. & Moss, R. J. 1980 *Cem. Concr. Res.* **10**, 665.
Golovastikov, N. I., Matveeva, R. G. & Belov, N. V. 1975 *Kristallografiya* **20**, 721.
Groves, G. W. 1981*a* *J. Mater. Sci.* **16**, 1063.
Groves, G. W. 1981*b* *Cem. Concr. Res.* **11**, 713.
Groves, G. W. 1982 *Cem. Concr. Res.* **12**, 619.
Hudson, K. E. & Groves, G. W. 1981 *Cem. Concr. Res.* **12**, 61.
Jeffery, J. W. 1952 *Acta Crystallogr.* **5**, 26.
Jennings, H. M., Dalgleish, B. J. & Pratt, P. L. 1981 *J. Am. ceram. Soc.* **64**, 567.
Taylor, H. F. W. 1964 In *The chemistry of cements* vol. 2, p. 367. London: Academic Press.

Discussion

J. P. Skalny (*Martin Marietta, Baltimore, U.S.A.*). First, is it possible that the differences in the morphology of C_3S with or without calcium chloride are due to different rates of C_3S hydration, resulting in different morphologies at equal time (three hours)? Second, is the observed 'porosity' due to damage by the electron beam?

G. W. Groves. Three hours with calcium chloride is equivalent to about six hours without and there is a marked difference between the morphologies after three hours with calcium chloride and six hours without. In answer to the second question, from my experience, no.

L. J. Parrott (*Cement and Concrete Association, Slough, U.K.*). Was the crack, shown in Dr Groves's micrograph of water hydrated tricalcium silicate, due to drying shrinkage of the calcium silicate gel?

G. W. Groves. This is perfectly possible.

K. Kendall (*I.C.I. New Science Group, Runcorn, Cheshire, U.K.*). Dr Groves showed electron micrographs of thinned samples of cement clinker that had been dipped in water and became covered with a layer of gelatinous material. Is this gelatinous layer the elusive membrane referred to by Dr Double in his paper? If so, is there any electron microscope evidence that the gel layer is disrupted by an osmotic pumping mechanism as suggested by the membrane model of cement hydration?

G. W. Groves. The answer to the first question is that I do not know and the answer to the second question is no.

Phil. Trans. R. Soc. Lond. A **310**, 85–92 (1983) [85]
Printed in Great Britain

Microanalytical studies (X-ray photoelectron spectrometry) of surface hydration reactions of cement compounds

By Micheline Regourd
Microstructures Department, C.E.R.I.L.H., Paris, France

X-ray photoelectron spectrometry (X.p.s.) measures the kinetic energy of electrons photoejected from a solid surface by soft X-rays. The kinetic energy of the photoelectrons can be related to the binding energy that these electrons had originally in the solid. X.p.s. is a rather new technique for studying cements. It has been used recently in the surface analysis of C_3S, C_2S, C_3A and blast-furnace slag grains during their hydration. Changes in chemical composition have been found as soon as the surface comes into contact with water, shown by a change in the shape, position and intensity of characteristic peaks like Ca_{2p}, Si_{2p}, O_{1s} and a reduction of characteristic ratios Ca/Si or Al/Si. A tentative interpretation of X.p.s. kinetic curves as a function of hydration time is presented.

1. Introduction

The initial hydration of cement compounds has not yet been completely clarified because of the complexity of the physico-chemical changes up to the end of the dormant period. Among the microanalytical techniques, the photoelectron spectrometry X.p.s. is a rather new one in the cement field but it has already been used for the study of the early hydration of cement compounds. X.p.s. measures the kinetic energy E_K of electrons photoejected from a solid surface. The kinetic energy is related to the binding energy E_B that these electrons had originally in the solid. X.p.s. is a surface analysis method as the depth of investigation is between 5 and 10 nm, and a quantitative method as it can be standardized on materials of known composition. This paper reports the modification of the chemical composition at the surface of cement compound grains that occurs as soon as they come into contact with water.

2. Study of the early hydration of tricalcium silicate

(i) *X.p.s. data*

A typical X.p.s. spectrum of C_3S shows the peaks Ca_{2p} (E_B = 346.9 eV), Si_{2p} (E_B = 101.2 eV), O_{1s} (E_B = 530.9 eV). The variation of Ca_{2p}/Si_{2p} X.p.s. ratios against time of hydration (figure 1) of C_3S pastes (w/s = 0.5) proves the importance of the perturbations on the surface at very early ages (Thomassin *et al.* 1979). The atomic Ca/Si ratio related to the X.p.s. Ca_{2p}/Si_{2p} ratio relative to a standard curve is, at any time of hydration, lower than in anhydrous C_3S. The binding energy and the shape of the Si_{2p} peak change within a very short period (table 1) (Regourd *et al.* 1980*a*). The increase of the O_{1s} binding energy from 532.9 eV, at t = 0, to 533.5 ± 0.2 eV, at t = 10 min, and simultaneously a broadening of the X.p.s. peak from 2.7 to 3.5 eV have also been observed (Ménétrier *et al.* 1979).

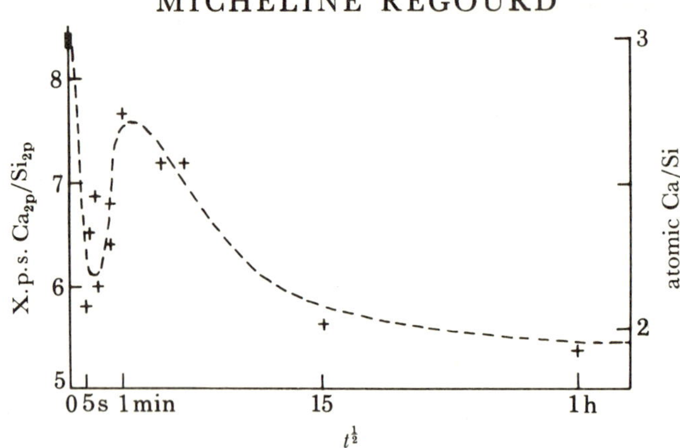

FIGURE 1. C_3S: variation of characteristic X.p.s. Ca_{2p}/Si_{2p} ratio and atomic Ca/Si ratio against (time of hydration)$^{\frac{1}{2}}$ (Thomassin *et al.* 1979).

TABLE 1. C_3S: DIFFERENCE BETWEEN BINDING ENERGIES OF X.P.S. PEAKS AND WIDTH OF THE Si_{2p} PEAK (REGOURD *et al.* 1980*a*)

time of hydration	$\delta(Ca_{2p}-Si_{2p})$/eV ($\pm$0.2 eV)	$\delta(Si_{2p}-O_{1s})$/eV ($\pm$0.2 eV)	(width of Si_{2p} peak)/eV ($\pm$0.1 eV)
0	247.7	429.7	2.5
5 s, 10 s, 15 s, 30 s, 1 min, 3 min, 5 min, 15 min, 1 h, 2 h	245.1	429.3	2.8
4 h	245.1	429.1	2.4

(ii) *Interpretation of X.p.s. data*

It is admitted that the beginning of the hydration is an instantaneous dissolution. A congruent dissolution could be assumed (Fuji *et al.* 1975; Barret *et al.* 1977), although an incongruent dissolution was also stated (Skalny *et al.* 1980).

The second stage, detectable in the first few seconds, corresponds to a drop of the Ca/Si ratio from 3 to 2.1 (figure 3). The cloud of points shows that the surface between 5 s and 1 min is a site of continuous exchange and is not in a state of equilibrium. The broadening of Si_{2p} and O_{1s} peaks indicates changes in the environment of electrons in the superficial layer. The higher binding energy of O_{1s} has been related to the existence of O—H bonds (Ménétrier *et al.* 1979).

A tentative quantitative interpretation of the X.p.s. data has been done according to the formula $I = \int_0^\infty \mathrm{d}I$, where I is the area of the photoelectron peak corresponding to a given energy and $\mathrm{d}I$ the contribution of a thickness $\mathrm{d}x$ to the depth x (Regourd *et al.* 1980*a*),

$$\mathrm{d}I = \sigma FSC\, \mathrm{e}^{-x/\lambda}\mathrm{d}x \qquad (1)$$

with σ the cross section of the sublayer irradiated by the X-ray beams, F the X-ray flux, C the concentration within the solid, and λ the mean free path of electrons, S a function of the apparatus. Considering a model of two layers, i.e. anhydrous C_3S and hydrated C_3S with an average Ca/Si close to 1 (Stein *et al.* 1977; Dent-Glasser *et al.* 1978) and assuming the thickness of the hydrated layer to be uniform and the mean free path of electrons Ca_{2p} and Si_{2p} to be the same in both layers, we obtain for the integration of (1)

$$\left.\begin{aligned} I_{\mathrm{Ca}} &= \sigma_{\mathrm{Ca}}\, FS_{\mathrm{Ca}}\left(\int_0^{d_1} C^1_{\mathrm{Ca}}\, \mathrm{e}^{-x/\lambda}\mathrm{d}x + \int_{d_1}^{\infty} C^2_{\mathrm{Ca}}\, \mathrm{e}^{-x/\lambda}\mathrm{d}x\right) \\ I_{\mathrm{Si}} &= \sigma_{\mathrm{Si}}\, FS_{\mathrm{Si}}\left(\int_0^{d_1} C^1_{\mathrm{Si}}\, \mathrm{e}^{-x/\lambda}\mathrm{d}x + \int_{d_1}^{\infty} C^2_{\mathrm{Si}}\, \mathrm{e}^{-x/\lambda}\mathrm{d}x\right) \end{aligned}\right\} \text{(1, hydrate; 2, } C_3S\text{).}$$

From the standard curve, for a value of $I_{Ca}/I_{Si} = 2.25$ at $t = 10$ s, the computed thickness d_1 of the hydrated layer is 0.8 nm.

The third stage manifests itself by an increase of the I_{Ca}/I_{Si} ratio in which the maximum is attained at the end of 1 min of hydration (figure 3). This could be explained by a chemisorption of Ca^{2+} ions on a surface enriched in silicon (Tadros *et al.* 1976). However, the Ca/Si of the hydrated layer would be lower than the observed maximum value of 2.7 because it includes a contribution from underlying C_3S. The calculation of I_{Ca}/I_{Si}, for Ca/Si close to 2, in the hydrated layer at $t = 1$ min gives a thickness of 0.8 nm, which strongly suggests that during the first stages of hydration the fixation of Ca^{2+} occurs from the solution only. The value of 0.8 nm is comparable to the thickness of the layer of C_3S dissolved after 1 min in the experiments of Fuji & Kondo (1975). The value of Ca/Si = 2 in the hydrate growing on C_3S seems to be confirmed by the same value obtained when the hydrated layer is thicker than 6 nm at $t > 15$ min. The Ca^{2+} ions can originate from two places, the solution and the inner part of the grain. These latter ions are released by the solid during the advancement of the interface hydrate, C_3S. The hydrated layer would develop and persist through the dormant period, up to the supersaturation of the solution.

After 4 h, important reorganizations take place in the hydrated layer, as evidenced by the change in the Si_{2p} peak. As already known, the end of the dormant period manifests itself by the nucleation of C—S—H with dimeric silicate ions $Si_2O_7^{6-}$ (Tamas *et al.* 1976) and $Ca(OH)_2$.

(iii) *Role of retarders*

The kinetic curve of C_3S hydrated with a solution containing NaF 10 g l^{-1} shows the instantaneous drop in Ca/Si ratio and a plateau at Ca/Si = 2 (figure 2), which is reached at 3 min (Regourd *et al.* 1980*b*). The variations of the binding energy and the shape of the Si_{2p} peak are the same as for C_3S hydration in distilled water.

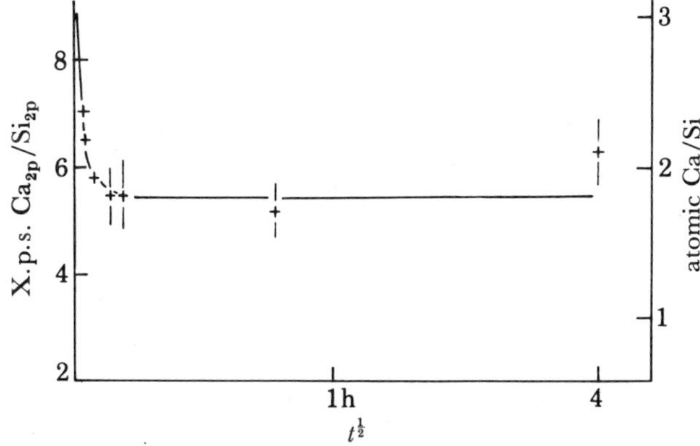

FIGURE 2. $C_3S + 1\%$ NaF: variation of characteristic X.p.s. Ca_{2p}/Si_{2p} ratio and atomic Ca/Si ratio against (time of hydration)$^{\frac{1}{2}}$ (Regourd *et al.* 1980).

The interpretation of the X.p.s. kinetic curve is not univocal. The lack of increase in the I_{Ca}/I_{Si} ratio could be related to the simultaneous formation of a hydrated layer, with Ca/Si $\ll 3$, and CaF_2 (Young *et al.* 1977), as the peak of the fluoride element is more intense than in NaF. Kinetic curves of C_3S hydrated in solutions containing amounts of NaF increasing from 0.001 % to 1 % show that the chemisorption of the Ca^{2+} ions on the surface enriched in silica is slowing down proportionally to the amount of NaF and disappears at 1 % NaF (Regourd *et al.* 1982). The dormant period is extended.

The same shape of kinetic curve as that of C_3S with NaF has been done for C_3S hydrated at 4 °C (Ménétrier *et al.* 1979). Low temperatures are well known to delay the setting of the cement pastes, but X.p.s. shows that if the first stage, the drop of Ca/Si ratio, remains unchanged, the superficial chemisorption of Ca^{2+} ions is missing, presumably in relation to the higher solubility of $Ca(OH)_2$ at 4 °C than at 20 °C.

3. Study of the dicalcium silicate hydration

The same characteristic X.p.s. peaks (Regourd *et al.* 1980*c*) and the same shape of kinetic curve (Thomassin *et al.* 1980; Jawed *et al.* 1980) as those of C_3S have been obtained for C_2S (figure 3), although the changes in surface composition were less pronounced and occurred later. The variations in the binding energies and the width of Si_{2p} peak have also been measured (table 2).

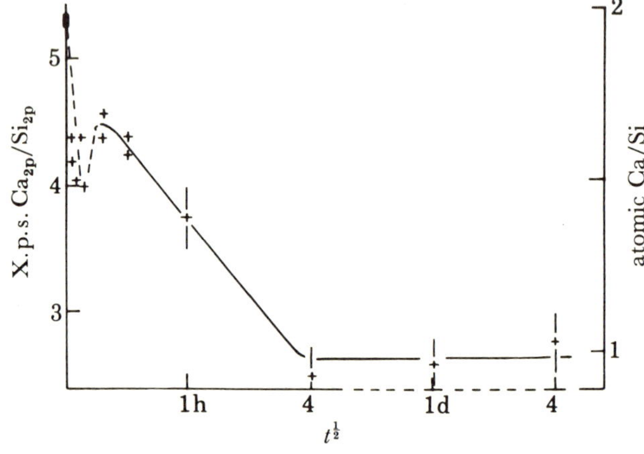

Figure 3. βC_2S: variation of characteristic X.p.s. Ca_{2p}/Si_{2p} ratio and atomic Ca/Si ratio against time of hydration (Thomassin *et al.* 1979).

Table 2. C_2S: difference between binding energies of Ca_{2p} and Si_{2p} peaks and width of the Si_{2p} peak (Thomassin *et al.* 1980)

time of hydration	$\delta(Ca_{2p}-Si_{2p})$/eV ($\pm$0.2 eV)	(width of Si_{2p} peak)/eV ($\pm$0.2 eV)
0	245.4	2.7
10 s, 15 s, 30 s, 5 min, 15 min	245·2	3.0
1 h	245.0	3.2
4 h, 24 h	244.6	3.4
4 d	244.9	3.4

The Ca/Si ratio at first drops by up to 1.5 at 1 min and the plateau at Ca/Si = 1 is reached at 4 h, remaining unchanged for up to 4 days in our study. Considering the same mechanism of hydration as for C_3S, i.e. a hydrated layer depleted in calcium and then a chemisorption of Ca^{2+} leading to Ca/Si = 1 during the dormant period, the calculated thickness is 0.8 nm at 5 min, 2 nm at 1 h and 6 nm at 4 h (Thomassin *et al.* 1980). Somewhat different results were obtained by Jawed *et al.* (1980) who, estimating the depth of their X.p.s. probing to be only 2 nm, used an indirect measurement of the hydrated layer by sputtering the surface of their pellets and then measuring the X.p.s. Ca/Si ratio. This evaluated depth was 5 nm after 15 s and 10 nm after 6 h.

4. Study of the early tricalcium aluminate hydration

The kinetic curve of C_3A hydration (figure 4) shows that the Ca/Al atomic ratio rises to about 2.0 at 2 sec, then decreases to about 1.7 at 1 min and stays there for up to 3 h (Jawed *et al.* 1982). The X.p.s. results do not prove the existence of an Al-rich superficial layer. The Ca/Al ratio close to 2.0 would suggest the formation of hexagonal C_4AH_x and the decrease to 1.7 would correspond to the formation of cubic C_3AH_x.

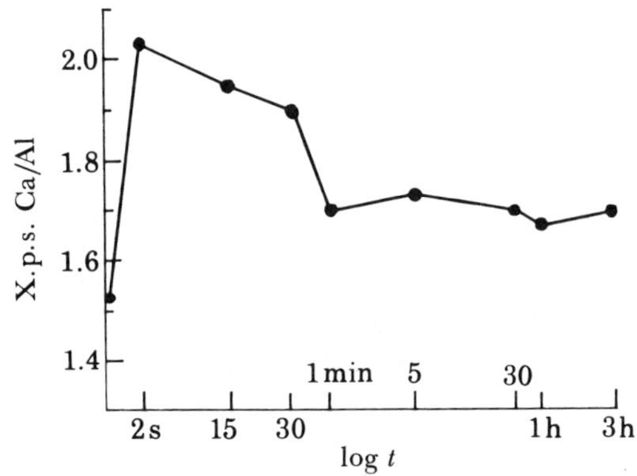

FIGURE 4. C_3A: variation of atomic Ca/Al ratio on surface as a function of hydration time (Jawed *et al.* 1982).

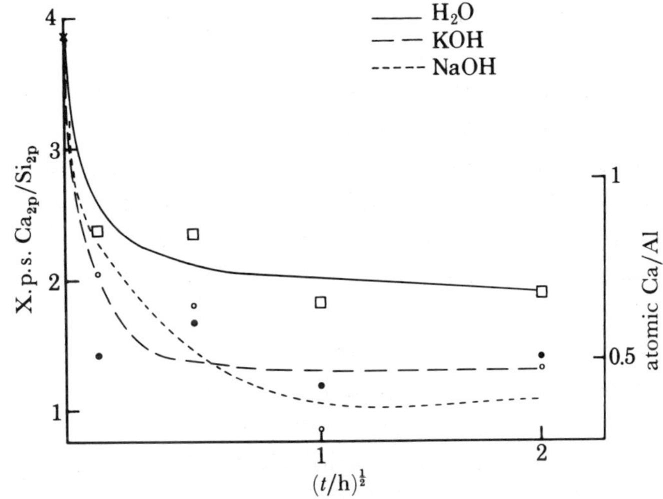

FIGURE 5. Granulated blast-furnace slag: variation of characteristic X.p.s. Ca_{2p}/Si_{2p} ratio and atomic Ca/Si ratio against (time of hydration)$^{\frac{1}{2}}$ (Regourd *et al.* 1981).

5. Study of the early hydration of blast-furnace slag by X.p.s.

Just as for C_3S and C_2S, granulated and pelletized slags react with water or alkaline solutions within minutes (Vernet *et al.* 1981). The Ca/Si ratio decreases rapidly at first, but after 1 min gradually reaches an essentially constant value (figure 5) (Regourd *et al.* 1981). After 15 min the different hydraulic slags tested, granulated or pelletized, gave the same Ca/Si ratio despite their different initial value. So, only the very first step of hydration, the incongruent dissolution (Dron *et al.* 1980), is related to the particular reactivity of the slag specimen. Sodium Na_{2s} (E_B = 64.5 eV) and potassium K_{2p} (E_B = 292 eV) are detected at the surface of slags hydrated in alkaline solutions, incorporated into the hydrated layer. The value of the Ca/Si ratio at the plateau is 0.8 in water and 0.5–0.6 in alkaline solutions. The trapping of alkalis has been confirmed later by the electron probe microanalysis. A very slight change was found to occur in the variation of Al_{2s}/Si_{2p} ratio (E_B = 120 eV for Al_{2s}) (figure 6). We could assume after the binding energies that Al and Si were still in tetrahedral coordination after having gone through the solution.

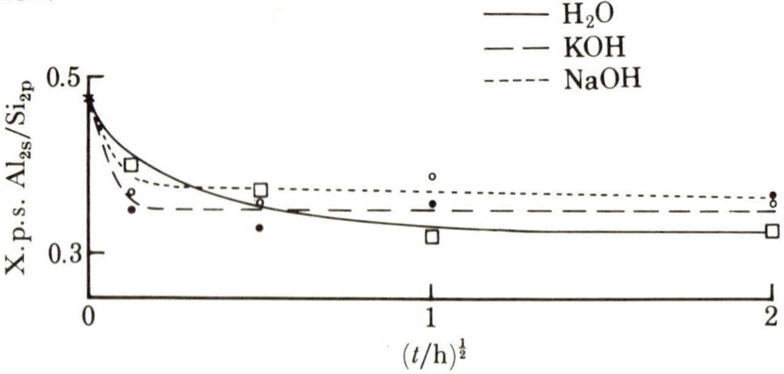

Figure 6. Granulated blast-furnace slag: variation of characteristic X.p.s. Al_{2s}/Si_{2p} ratio against (time of hydration)$^{\frac{1}{2}}$ (Regourd *et al.* 1981).

6. Conclusion

X.p.s. appears to be a very sensitive method in the study of early hydration in cement compounds. Surface transformations of C_3S, C_2S, C_3A and blast-furnace slag grains are observed from the first few seconds of their contact with water. These modifications are related to a change of the environment of Si and O atoms and a variation of the characteristic ratios, i.e. a decrease for Ca/Si and Al/Si in C_3S, C_2S and slags, an increase for Ca/Al in C_3A. So, the first drop in Ca/Si ratio in C_3S pastes does not indicate the formation of a first hydrate with Ca/Si = 3 and there is no Al-rich superficial layer in the C_3A early hydration.

The interpretation of the variation of X.p.s. characteristic ratios with time is not always easy and univocal, as the depth of investigations is of the order of only 6 nm. Results reported here have assumed a double layer model with a uniform hydrated layer on an anhydrous substratum. The computed thickness has required an average composition of the different hydrates (Ca/Si $\approx$ 1 then Ca/Si $\approx$ 2 for C_3S). More research is needed to explain the action of retarders like NaF in the early hydration of C_3S. Lastly, microanalytical X.p.s. results of surface analysis need to be compared to data from other electron optical techniques and s.t.e.m. in particular.

References

Barret, P., Ménétrier, D. & Bertrandie, D. 1977 *Rev. Int. Hautes Temp. & Réfrac.* **14**, 127.

Dent-Glasser, L. S., Lachowski, E. E., Mohan, K. & Taylor, H. F. W. 1978 *Cem. Concr. Res.* **8**, 733.

Dron, R. & Brivot, F. 1980 *7th Int. Congress on the chemistry of Cements, Paris*, vol. II, III – 134.

Fuji, K. & Kondo, W. 1975 *J. Ceram. Ass. Japan* **83**, 214.

Jawed, I., Ménétrier, D. & Skalny, J. 1980 *7th Int. Congress on the chemistry of Cements, Paris*, vol. II, III – 182.

Jawed, I. & Skalny, J. 1982 *Int. Sem. on Calcium Aluminates, Torino*, vol. II, 3.

Ménétrier, D., Jawed, I., Sun, T. S. & Skalny, J. 1979 *Cem. Concr. Res.* **9**, 473.

Regourd, M., Thomassin, J. H., Baillif, P. & Touray, J. C. 1980*a* *Cem. Concr. Res.* **10**, 223.

Regourd, M., Thomassin, J. H., Baillif, P. & Touray, J. C. 1980*b* In *7th Int. Congress on the chemistry of Cements, Paris*, vol. II, II – 123.

Regourd, M., Thomassin, J. H., Baillif, P. & Touray, J. C. 1980*c* In *9th Int. Symp. on the Reactivity of Solids, Cracow, Poland*, VI – 39, 769.

Regourd, M., Thomassin, J. H., Baillif, P. & Touray, J. C. 1981 83rd Am. Ceram. Soc. meeting, Washington. To be published.

Regourd, M., Thomassin, J. H., Baillif, P. & Touray, J. C. 1982 84th Am. Ceram. Soc. meeting, Cincinnati. *Cem. Concr. Res.* **13**, 549, 1983.

Skalny, J. & Young, F. Y. 1980 In *7th Int. Congress on the chemistry of Cements, Paris*, vol. I, II – 1/3. Paris: Éditions Septima.

Stein, H. N. 1977 *Ind. Cemento*, **74**, 3.

Tadros, M., Skalny, J. & Kalyoncu, R. 1976 *J. Am. ceram. Soc.* **59**, 344.

Tamàs, F. D., Sarkar, A. K. & Roy, D. M. 1976 In *Hydraulic cement pastes: their structure and their properties*, p. 55. Slough: Cement and Concrete Research Association.

Thomassin, J. H., Regourd, M., Baillif, P. & Touray, J. C. 1979 *C. r. hebd. Séanc. Acad. Sci., Paris*, C **288**, 93.

Thomassin, J. H., Regourd, M., Baillif, P. & Touray, J. C. 1980 *C. r. hebd. Séanc. Acad. Sci., Paris*, C **290**, 1.

Vernet, C., Demoulirn, E., Gourdin, P. & Hawthorn, F. 1980 In *7th Int. Congress on the chemistry of Cements, Paris.* vol. II, III – 128. Paris: Éditions Septima.

Young, J. F., Tong, H. & Berger, R. L. 1977 *J. Am. ceram. Soc.* **60**, 193.

Discussion

C. Defossé (*Dowell-Schlumberger Research and Development, Z.I. Molina La Chazotte BP* 90, *F.* 42003 *Saint-Etienne, Cedex*). Concerning drying treatment, can Dr Regourd exclude any possibility of having species in solution precipitating on the surface of the grain at the very moment of the quenching? For instance, fluoride ions or CaF_2 in suspension could precipitate on the surface and account for the F_{1s} peak observed.

M. Regourd. We cannot exclude any possibility of CaF_2 precipitating from the solution on the C_3S surface. We only observed that the intensity of the F_{1s} peak was higher in $C_3S + 1\,\%$ NaF than in NaF, after several minutes of hydration.

S. A. Jefferis (*Civil Engineering Department, King's College London, Strand, London, WC2R 2LS, U.K.*). I was interested to see Dr Regourd's data on the changes in surface chemistry that occur during the first seconds and minutes after cement components are wetted and to hear that she feels that this could affect mix rheology.

Such rapid reactions could explain a phenomenon that I have observed when ordinary Portland cement is mixed with water containing a chloride based accelerator. The rheology and subsequent setting times of the particular mixes appeared to depend very much on the level of shear during the first seconds of mixing and to be largely independent of any subsequent shearing. Low shear mixing may not disperse any clusters of anhydrous grains present in the dry cement. If such clusters are very rapidly tied together by surface reactions, subsequent high shear may not be able to disperse them. Whereas cements initially subjected to high shear may be more fully dispersed and thus present more active surface area with its consequent effects on rheology and setting time.

Does Dr Regourd have any data or hypotheses on how surface reactions affect rheology?

M. Regourd. Surface reactions affecting rheology have been shown through Zeta potential measurements. The early hydration reactions leading to the dissolution of ionic species modify the superficial electrical charge on cement grains then disperse the flocculi by mutual repulsion.

Calcium chloride is an accelerator for C_3S, it forms chloro-aluminates with C_3A and leads to a more dispersed calcium hydroxide crystallization. So its action may be effective in the early hydration, i.e. it may play a role in the precipitation of hydrates and the modification of the double electrical layer.

J. Bensted (*Blue Circle Technical Research Division, Greenhithe, U.K.*). Has Dr Regourd looked at the different types of C_3S polymorphs by X.p.s. or e.s.c.a to see whether there are differences in the early reactivity.

M. Regourd. Pure triclinic C_3S was used in this study. We have not yet studied C_3S solid solutions.

Phil. Trans. R. Soc. Lond. A **310**, 93–103 (1983) [93]
Printed in Great Britain

Electron microscope studies of Portland cement microstructures during setting and hardening

By P. L. Pratt and A. Ghose†
Department of Metallurgy and Materials Science, Imperial College, Prince Consort Road, London SW7 2BP, U.K.

[Plates 1 and 2]

The rates of hydration of the individual phases in Portland cement have been related both to the disappearance of these phases from the microstructure and to the appearance of their products of hydration.

After 12–15 h hydration a shell of product is formed containing short rods of AFt, an infilling of C–S–H and ettringite rods 1–2 μm long from the hydration of C_3A. Later hydration takes the form of Hadley grains, by dissolution of the core from inside the shell and precipitation of product either inside the shell or in between the shells. The network of shells forms the cohesive element in the system after 10–12 h, with the coherent inner product filling up the space progressively inside the shells but not being attached to them. The implication of these loose fillings for the tensile strength and permeability are discussed.

The onset of rapid hydration of C_4AF has been identified by quantitative X-ray diffraction with a new peak found by conduction calorimetry; in the microstructure, fibres of AFt phase, 5 μm or more in length, appear suddenly from the C_4AF and these are a familiar feature of o.p.c. pastes after 2 or 3 days hydration. After about 5 days at 20 °C the local availability of water is reduced and the rate of hydration of all the phases is reduced.

1. Introduction

In previous papers on C_3S (Jennings *et al.* 1981) and on Portland cement (Dalgleish *et al.* 1982*a*), electron microscope techniques were combined with a variety of methods of specimen preparation to examine the development of the microstructure with time of hydration. Differences between early, middle and late products of hydration were identified and these differences were related to the stages of hydration by using an isothermal calorimeter. Despite early suggestions that the various clinker compounds in normal Portland cements have the same partial rate of hydration (Brunauer 1957), subsequent work has shown that this is not the case. Locher (1981) showed clearly that the induction period for C_3S in Portland cement ends well before that for C_3A and this is confirmed by the X-ray data of Osbaeck & Jons (1980). In this paper the rates of hydration of the individual phases in Portland cement are related both to the disappearance of those phases from the microstructure and to the appearance of the products of their hydration. Thereby new features of the mechanisms of hydration of cement are revealed. Since the earliest stages of hydration have received attention recently (Dalgleish *et al.* 1981; Dalgleish *et al.* 1982*b*), this paper will concentrate on hydration from 24 h and onwards.

† Present address: Department of Civil Engineering, University of Illinois at Urbana-Champaign, Urbana, Illinois 61801, U.S.A.

2. Experimental methods

The reactions occurring in the cement pastes and the resulting development of microstructure were studied with conduction calorimetry, quantitative X-ray diffractometry, and scanning electron microscopy. Northfleet O.P.C. was mixed with distilled water at a water/cement ratio of 0.5. For calorimetric studies, a Wexham Instruments isothermal calorimeter was used at 20 °C. The specimens were mixed by kneading for 5 min in a polythene bag outside the calorimeter so that the first heat peak was missed. For s.e.m. observations, the pastes were cast into small cylindrical moulds, sealed and left to cure at room temperature. At appropriate times the moulds were split open, the specimens fractured in bending while still wet and immediately flooded with isopropanol to stop the hydration reactions. After about 15 min, they were dried in an oven for about 2 h, mounted on stubs, coated with 25–30 nm of gold and stored in a vacuum desiccator until the s.e.m. was available. For q.X.r.d., part of the s.e.m. specimen was powdered in a micronizing mill and measurements and analysis were done at the Cement and Concrete Association, Wexham Springs.

Table 1. Cement composition (%)

chemical		Bogue		q.X.r.d.	
SiO_2	20.3	C_3S	59.0	C_3S	67.0
Al_2O_3	5.2	C_2S	13.7	β-C_2S	13.0
Fe_2O_3	3.1			α-C_2S	2.0
CaO (total)	64.7	C_3A	8.5	C_3A	5.0
Na_2O	0.19	C_4AF	9.4	C_4AF	6.0
MgO	1.1	$\bar{K}$ (total)	0.63	anhydrite	2.0
K_2O	0.34			hemihydrate	2.5
SO_3	2.8				
free lime	0.7				

The composition of the Northfleet cement is given in table 1. The difference between the Bogue analysis and that measured by q.X.r.d. is significant especially for the aluminate phases. The specific surface area was 341 $m^2 kg^{-1}$. This standard production Northfleet O.P.C. is similar to that used before (Dalgleish *et al.* 1981; Dalgleish *et al.* 1982*a*) and is the same as cement B in the paper by Dalgleish *et al.* (1982*b*). (In the latter publication the Bogue composition for cement B is reported incorrectly although the oxide analysis is correct.) Particle size analysis of the Northfleet O.P.C. with a Coulter counter showed up to 30 % (by volume) of particles greater than 30 μm and up to 3 % greater than 90 μm. Many of the large particles were polymineralic, but the average particle size for those that were largely C_3S, or alite, was 20–25 μm.

3. Experimental results

(*a*) *Degree of hydration of cement phases*

The rate of heat evolution with time for the Northfleet O.P.C. is shown in figure 1. As we have noticed previously with this cement (Ghose & Pratt 1981), in addition to peak 2, which is due to the hydration of C_3S forming middle product C–S–H/CH, and peak 3, which is due to the hydration of C_3A forming rods of ettringite, a distinct peak 4 can be seen. This peak is accelerated and enhanced by the replacement of 30 % of the O.P.C. by fly ash. This peak 4 has now been identified by q.X.r.d. as marking the onset of rapid hydration of C_4AF. During peak 4,

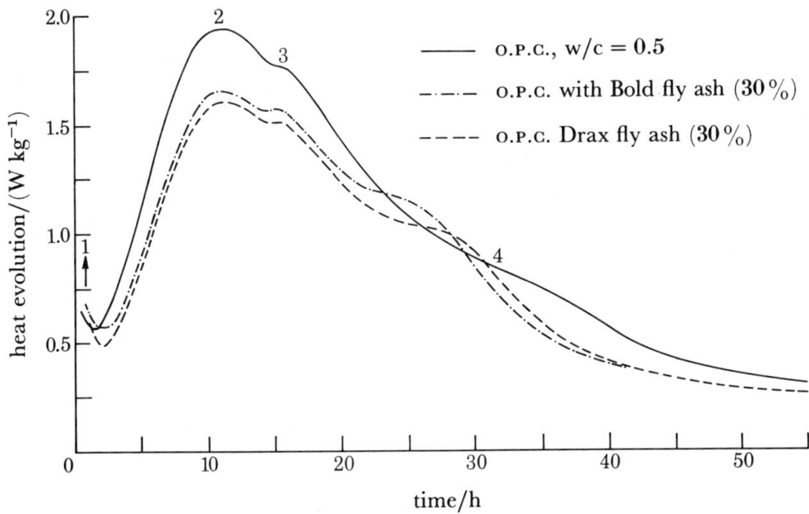

FIGURE 1. The rate of evolution of heat at 20 °C for Northfleet o.p.c. and o.p.c. with fly ash blends at w/s ratio 0.5.

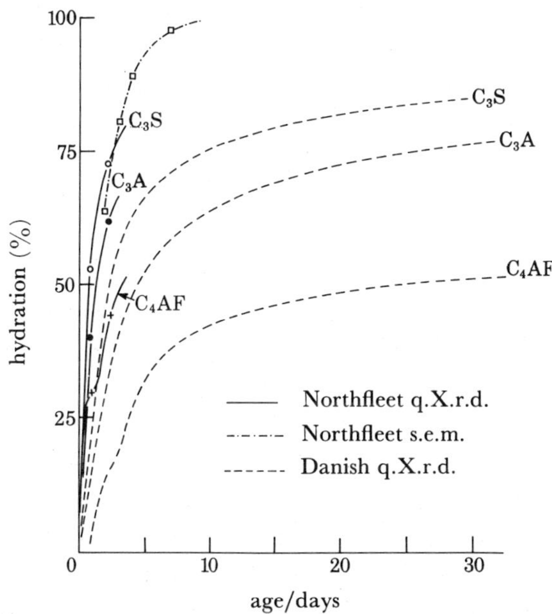

FIGURE 2. The degree of hydration as a function of time for the individual phases in Northfleet cement. Data for Danish cement at w/c ratio 0.44, after Jons & Osbaeck (1982).

hydration of some 10 % of the C_4AF in neat cement paste takes place, and 20 % in the cement/fly ash mixtures. Fukuhara *et al.* (1981) have measured the heat of hydration for C_4AF and gypsum to form AFt as 724 J/g. From this figure, with the percentage of C_4AF that hydrated during peak 4 determined by q.X.r.d., the total heat evolved during peak 4 can be accounted for satisfactorily (Halse & Pratt 1983).

The recent X-ray data of Osbaeck & Jons (1980) and Jons & Osbaeck (1982) on the hydration of cement phases is more complete than ours and part of their data is plotted together with ours in figure 2. In both cements the rates of hydration of the phases C_3S, C_3A and C_4AF differ significantly; C_3S hydrates faster than C_3A, whereas C_4AF lags noticeably behind. The data for C_2S are less reliable but by 28 days some 25 % has hydrated. The Danish cement resembles the Northfleet O.P.C. with a $\bar{\bar{K}}$ (total) of 0.75 % compared with our 0.63 %, although by Bogue analysis it contains 51 % C_3S and 27 % C_2S, compared with our 59 % and 13.7 % by Bogue.

(*b*) *Development of microstructure*

As with all the cements we have examined (Dalgleish *et al.* 1982*a*, *b*), the early formation of very short rods of AFt, 0.2 μm long, on the surface of the particles is followed at the end of the induction period for C_3S by the build-up of a thickening shell of C–S–H around the cement grains engulfing the AFt. After 12 h hydration this shell is between 0.5 and 1 μm thick and the cohesion between neighbouring shells is strong enough to enable the paste to be broken into two pieces without collapsing to powder. The fracture path runs through the shell in many places, revealing the formation of Hadley grains with a hollow space between the shell and the hydrating core. Longer rods of ettringite appear at about this time (10–12 h) associated with the end of the induction period for C_3A, peak 3 in figure 1. Some of these are clustered in the form of small spherulites but most appear as outgrowths from the shell of the Hadley grains with a length of 1–2 μm.

After 24 h hydration, the Northfleet O.P.C. shows many Hadley grains, figure 3, plate 1, with the characteristic spiky shell of ettringite rods and type I C–S–H. Most of the top of this micrograph shows the inner surface of a Hadley shell some 25 μm across with remnants of the hydrating core, which has been pulled away. In this micrograph there are a further nine Hadley grains, which can be seen by the practised eye; some like that in the centre of the figure have a lamellar core, which is believed to be from C_3A, and others like that at the top of the figure come from C_3S. Large crystals of CH grow up to 50 μm in size in the original water space and in this form they can be seen even after 4–5 days where the fracture path is inter-granular.

After 2 days, the induction period for C_4AF has ended, leading to peak 4, and a new and coarser morphology of AFt fibres appears, figure 4, typically 5 μm in length or more. This sudden appearance is even more noticeable with fly ash added to the cement, as would be expected from figure 1. The earlier rods of ettringite from the C_3A, 1–2 μm in length, are still visible in the background and the new long fibres of AFt are recognized as a familiar feature of S.E.M. micrographs of O.P.C. pastes after 2 or 3 days hydration. X-ray analysis shows that these fibres contain significant quantities of Al, $\bar{S}$ and Fe as well as Ca. In some Hadley grains the core appears to have contained both C_3A and C_4AF, from which the Ca, the Al, and only part of the Fe has dissolved, leaving amorphous iron-rich regions (Fukuhara *et al.* 1980). At this age the shells are starting to thicken by deposition of C–S–H and CH inwards. In figure 4 this thickening of the shell can be seen around a polymineralic Hadley grain. The upper part of the grain, largely dissolved, resembles that associated with C_3A in figure 3, whereas the lower

---

DESCRIPTION OF PLATE 1

FIGURE 3. 1 day hydration. Hadley grains, with ettringite rods.
FIGURE 4. 2 days hydration. Hadley grains with coarse AFt fibres.
FIGURE 5. 3 days hydration. C_3S core.
FIGURE 6. 4 days hydration. C_3S core.
FIGURE 7. 4 days hydration. α'-C_2S core.
FIGURE 8. 4 days hydration. β-C_2S emerging in C_3S core.

FIGURES 3–8. For description see opposite.

(*Facing p.* 96)

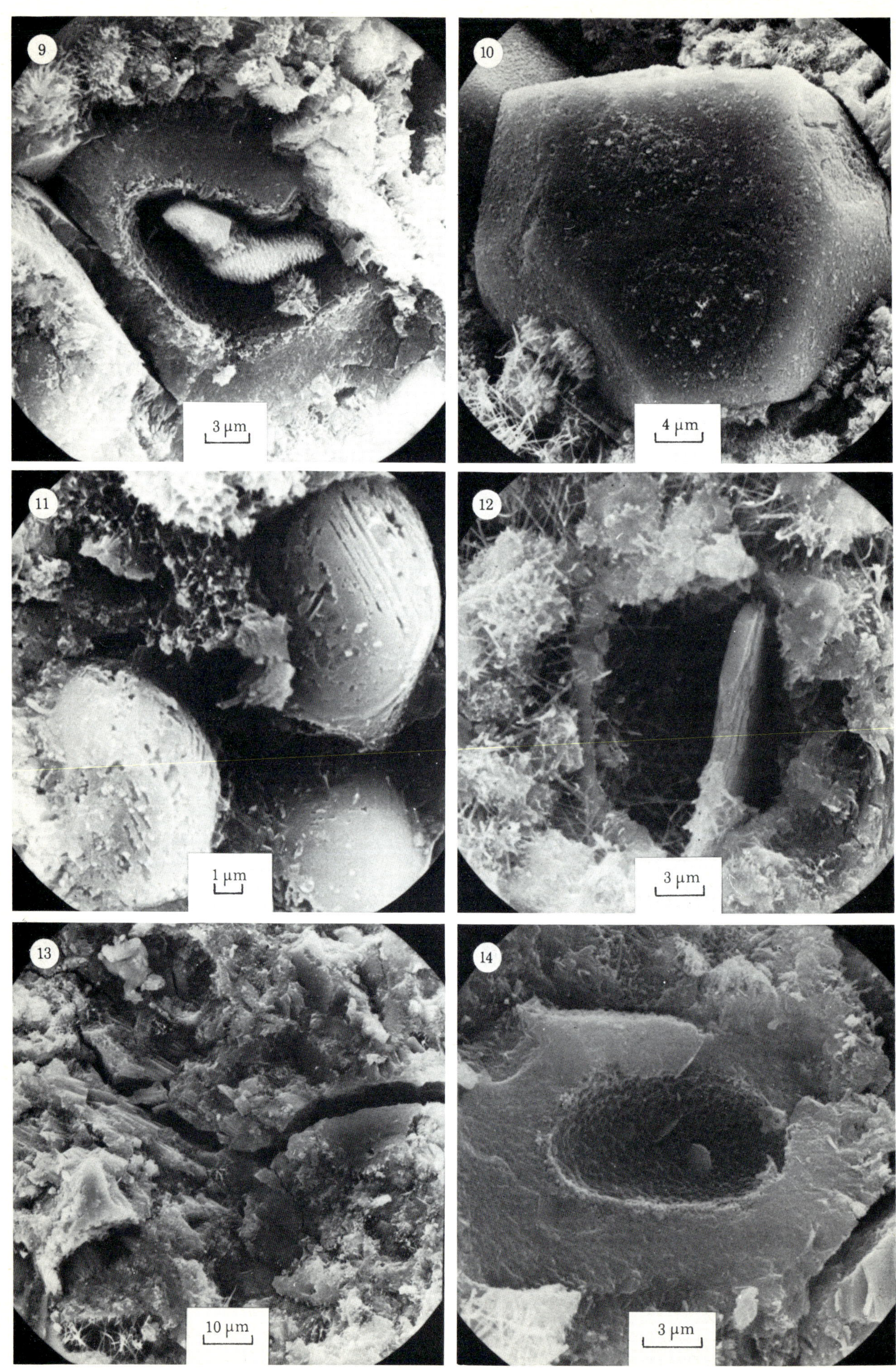

FIGURES 9–14. For description see opposite.

has the serrated appearance of the C_3S core in figure 5. After 3 days the shells have thickened to 2–3 μm, the C_3S cores are dissolved away substantially, figure 5, sometimes revealing spheres of β-C_2S, while the occasional large particle of C_4AF has continued to dissolve. At this time squat hexagonal platelets can be seen inside some of the shells although their occurrence is rare. They are believed to be the α′-C_2S, which makes up 2 % of this O.P.C. By this time the inward growth of the C–S–H/CH into the centre of the Hadley grain is becoming more noticeable and after 4 days, especially around C_3S cores, figure 6, the shell thickness can approach 5 μm. The smaller C_3S Hadley grains are by now almost filled with flaky C–S–H/CH whereas those that contain α′-C_2S cores, figure 7, remain relatively empty. Inside this particular shell, plates of AFm are growing inwards together with larger plates of CH. Outside the shell, however, both fine rods of ettringite and coarse fibres of AFt remain stable, suggesting that the shortage of gypsum is confined to the region within the shell. The C–S–H outside the shell appears to have dried into a type II morphology and this is fairly typical of the outer C–S–H at 4 days in this cement. The inner C–S–H/CH, especially in figure 8, has a characteristically porous appearance, composed of flakes or flattened equant grains growing together. Inside this shell, small rounded particles of β-C_2S are emerging from the dissolving C_3S.

After 7 days, the C_3S cores revealed by fracture continue to dissolve and the inner products continue to fill up the empty space within the shell around them, figure 9, plate 2. Depending upon the original particle size, the thickness of the shell can exceed 10 μm, so that particles originally 20 μm across could be filled with inner product. Depending upon the fracture path, the inner product can appear to take up the morphology of the original particle in the O.P.C., figure 10, in this case an angular alite grain overlying its neighbour. The mottled inner product is now coherent in itself but, in the dried state, it is detached from the outer Hadley shell save for a few patches of contact. In Hadley grains containing C_3S and β-C_2S, the β-C_2S starts to etch after 7 days, although the access of water is limited by the coherence of the inner product surrounding it. Between the Hadley grains the intergrowths of massive CH crystals are filling up the structure, although long fibres of AFt and spherulites can still be seen. The crack path has become flatter, with a substantial proportion of intra- rather than inter-granular fracture, and stepped cleavage planes through the massive CH crystals become a prominent feature. After 14 days this transparticle fracture becomes even more noticeable and at the same time the dissolution of C_2S, figure 11, is well under way.

After 28 days, some hollow shells remain, particularly those containing α′-C_2S, figure 12, and β-C_2S, both of which have hydrated now by some 25 %. After 2 months, the overall porosity is reduced and transparticle fracture is predominant, with large areas of cleavage through CH, figure 13. Outlines of former Hadley grains can be recognized by the structure of the outer shell and the fibres of AFt remaining in the unfilled interstices. Coherent inner product can be recognized on the fracture surface, figure 14, within which all the C_3S has

---

Description of plate 2

Figure 9. 7 days hydration. C_3S core and mottled inner product.
Figure 10. 7 days hydration. Inner product detached from shell.
Figure 11. 14 days hydration. β-C_2S in core.
Figure 12. 1 month hydration. α′-C_2S in core.
Figure 13. 2 months hydration. Transparticle fracture.
Figure 14. 2 months hydration. Detached inner product.

hydrated, leaving rounded cavities with a few small platelets of CH; this inner product has cracked away from the original outer shell. In other areas of figure 13, long rods of AFt can still be seen and these appear to be a permanent feature of the microstructure, at least up to 8 months. Between 2 and 4 months some of the rounded cavities in the coherent inner product fill slowly with a different form of C–S–H presumably coming from the hydration of C_2S.

4. Discussion

(a) *Rates of hydration of cement phases*

(i) C_4AF

The addition of 30% fly ash accelerated and enhanced peak 4 in the calorimeter curve in figure 1 and, for these blended pastes also, q.X.r.d. confirmed that the peak was due to the onset of rapid hydration of C_4AF. The pozzolanic reaction with CH is known to start in this blend after some 24 h hydration (Halse & Pratt 1983) and this provides the clue to understanding the enhancement of peak 4. Fukuhara *et al.* (1980) measured the rate of heat evolution for the hydration of C_4AF with additions of gypsum and lime, showing that the more lime was added the more the main hydration peak was retarded. The pozzolanic reaction of the fly ash reduces the quantity of CH in the hydrating system by forming C–S–H and consequently both accelerates and enhances the hydration of C_4AF. Similar effects would be expected with slag cements and with the addition of silica fume; a large early peak 4 can be seen in both (Goult & Pratt 1983). This same explanation can account for the different rates of hydration of C_4AF in cements of different alkali content (Osbaeck & Jons 1980). Increased alkali content accelerates the hydration of C_3S in the first few days, thereby increasing the quantity of CH in the system. The retarding effect on the hydration of C_4AF is very obvious from their data and it would be interesting to confirm this hypothesis by microscopic observation.

(ii) C_3S *and* C_3A

Whereas Mindess & Young (1981) in their recent textbook suggest that the hydration of C_3A in Portland cement paste is faster than that of C_3S, the data in figure 2 show that this is not so, at least for the Danish and British cements. Locher (1981) showed that for German Portland cement the induction period for C_3S ends well before that for C_3A so long as the cement contains gypsum, while the hydration of C_3S is the faster of the two. This agrees with our interpretation of peak 3 as the onset of hydration of C_3A giving rods of ettringite 1–2 μm in length, outside the Hadley shells. Fine particles of C_3A, and also of C_4AF, react during the first few minutes in peak 1 to form the amorphous gel layer with short rods of AFt 0.2 μm in length, seen in the microscope, but thereafter C_3S starts earlier and is the faster to hydrate in all three European cements. It would be interesting to know why the American cement referred to by Mindess & Young appears to behave differently.

(iii) C_3S, C_3A *and* C_4AF

Taplin (1959) measured the effect of w/c ratio upon the overall rate of hydration, assuming that all the cement phases hydrate at the same rate. He found a marked decrease in rate, at increasing degrees of hydration with increasing w/c, and this he ascribed to the filling up of available water space with hydration products. While allowances have to be made for the distribution of the sizes of the hydrating particles (Taplin 1969), at first sight the rates of hydration of C_3S, C_3A and C_4AF in the Danish cement in figure 2 all decrease markedly after about 5 days, the aluminate phases more sharply than C_3S. Rather than suggesting the filling of space with hydration product, this decrease in hydration rate suggests that the local availability of water to the anhydrous phases becomes reduced at this time.

(*b*) *Degree of hydration and development of microstructure*

The degree of hydration of the various phases measured by q.X.r.d. in figure 2 represents the average degree of hydration for the paste as a whole. At a particular time the smallest particles of one phase may be fully hydrated while the largest will remain largely unhydrated. Only the mean particle size will approximate to the rate of hydration shown in figure 2. For C_3S, or better alite, particles, figures 5, 6, 9 and 14 represent the sequence of hydration from 2 days to 2 months for particles originally about 20 μm in diameter. The volume of the alite cores remaining can be estimated crudely from these micrographs and these particles appear to be hydrating faster than the Danish cement, in agreement with our X-ray data in figure 2.

For C_3A the picture is less clear because identification of the C_3A particles is less obvious. The lamellar regions seen inside the Hadley shells in figures 3, 4 and 8 could be remnant C_3A, or alternatively they could be plates of monosulphate or hexagonal C_4AH_{13} where the SO_4^{2-} ion had difficulty in penetrating. Further work is needed to confirm the identity and the rate of hydration of these phases.

The work of Fukuhara *et al.* (1980) on the hydration of C_4AF makes it clear that the fibrous product AFt contains much less iron than the C_4AF from which it came. They suggested that an amorphous relict phase, rich in iron, with the same shape as the original C_4AF particle was left after dissolution of the material needed to form the AFt. Striated regions within the shell are a common feature of Hadley grains in cement after 2 days and their appearance coincides with that of the long coarse AFt fibres containing iron.

The hydration of α'-C_2S is somewhat faster than that of β-C_2S (Jelenić-Bezjak 1983) and the squat hexagonal platelets in figures 7 and 12 are believed to represent this phase. Some dissolution has occurred by 4 days but the rate of hydration is slow. Unlike the C_3S, these Hadley shells appear to remain largely unfilled as late as 1 month, when perhaps 25–30 % of the phase should have hydrated. By contrast, β-C_2S starts dissolving after 7 days but is frequently found inside Hadley grains in which C–S–H from the adjacent C_3S, figure 11, has already surrounded it. Under these circumstances, unlike α'-C_2S, the hydration is starting only when the free access of water has already been reduced, so that the rate is bound to be slow.

(*c*) *Mechanisms of hydration*

This microstructural study suggests that the presence of the interstitial aluminate phases in Portland cement determines the mechanisms of hydration. The principal difference between the hydration of O.P.C. and that of pure C_3S is the early formation of a shell around the cement grains in the former and the subsequent dissolution of the core from within this shell. The network of shells forms the coherent element in the hydrating cement system after 10–12 h and the space left both between the shells and within the shells is filled progressively by a variety of products.

(i) *Formation of the shell*

The skeleton of the hydration shell appears to be the short rods of AFt formed during peak 1 from the dissolution of C_3A and C_4AF. The infilling of this skeleton with C–S–H and some CH starts at the end of the induction period for C_3S and is almost completed by the top of peak 3 when the first Hadley grains were found. At the same time, renewed hydration of C_3A gives rise to rods of ettringite, 1–2 μm in length, bursting out from the shells and leaving empty space behind them. Recent experiments by Goult & Pratt (1983) show that hollow shells can form during the hydration of cements containing C_4AF and only very little C_3A. By q.X.r.d., Jons & Osbaeck (1982) found no ettringite after 1 day in a similar cement that contained 18 % C_4AF and no C_3A. After 3 days, ettringite, or more probably AFt, was present and no conversion to monosulphate was found by 28 days.

(ii) *Filling of space between the shells*

A major contribution to the filling of the original water space between cement particles after 12 h comes from the large crystals of CH. The corresponding C–S–H appears to be retained largely in the shell as infilling. The onset of hydration of C_3A covering the outer surface of the shells with ettringite rods, 1–2 μm long, figure 3, makes a further significant contribution, at the expense of creating space inside the shells. The onset of hydration of C_4AF after about 30 h gives rise to long coarse fibres of AFt, many of which are associated with spherulites (Dalgleish *et al.* 1982). These arise from the large number of small particles of the interstitial phase of the cement that contain C_4AF. The formation of monosulphate starts inside the shells as early as 24 h, where the supply of gypsum is limited. For their Danish cement, Jons & Osbaeck (1982) have shown that equal amounts of ettringite and monosulphate are present after 3 days, some of the monosulphate having formed by conversion of the ettringite. Although some ettringite rods and AFt fibres remain in open pores as late as 8 months, many must be converted to monosulphate as the hydration products continue to fill the original water space between the particles. The release of water as this conversion takes place enables the hydration to continue, although this water is released outside the original Hadley shells rather than inside, where the bulk of the unhydrated material remains.

(iii) *Filling of space within the shells*

Part of the difficulty of accepting the formation of Hadley grains as the principal mechanism of hydration from the top of peak 2 onwards has been the problem of the density of hydrated cement paste, if it is indeed full of hollow shells. The filling of shells containing C_3S, i.e. the majority of the shells in the cement, can now be seen to take place by dissolution of the core and precipitation of an intimate mixture of C–S–H/CH on the inside of the shell, figures 6, 9, 10 and 14. This inner product takes up the form of the original particle, figure 10, and protects the core, effectively cutting it off from the original water space after 5–6 days. This intimate mixture of product should have a C/S ratio reflecting the composition of the material within the core, approaching 3 if the core was largely C_3S and all the products were retained within the shell; Diamond (1976) has found a C/S ratio 2.96 for this inner product. It forms the greater part of the products of hydration of cement paste and can be recognized easily on the fracture surface of mature pastes. Often there is sufficient C_3S inside the shell to fill it with this product, but sometimes, as in figure 14, there is not. These cavities may be filled subsequently with product of a different texture coming either from the larger C_3S grains or from C_2S.

(iv) *Types of* C–S–H *in hydrated* O.P.C.

On the dried s.e.m. specimens, rolled-up foils and fibres of type I C–S–H (Diamond 1976), which start to form during the first few hours, remain visible for 3–4 days on the outer surfaces of the Hadley grains. Further hydration of C_3S leads to infilling at the roots of the fibres with the appearance of reticular type II C–S–H after 3 days. These are both artefacts of drying, reflecting the way in which the wet outer layers of C–S–H shrink and thus reduce their surface area. Diamond (1976) has designated the inner product morphology as type IV C–S–H, identifying it by its dimpled appearance with either regular pores or close packed equant grains about 0.1 μm in size. He suggested that recognizable 'inner product' will be lacking to the extent that the cement dissolves from within the Hadley shell. We have seen that dissolution of C_3S is followed by precipitation of C–S–H/CH within the shell. To this extent we believe that type IV C–S–H is the most common product in our cements. Type III C–S–H is more difficult to identify; some of Diamond's micrographs appear to refer to infilling in the Hadley shell, some to the core of spherulites and some to the products of hydration of small particles of alite.

(*d*) *Implications for strength and durability*

A number of implications for the strength, durability and permeability of hardened cement pastes arise from these observations of the microstructure and the modes of hydration. Perhaps the most important relates to the way in which the coherent type IV C–S–H/CH structure develops inside the original Hadley shell. In figure 10, at 7 days, and in figure 14, after 2 months, this structure in the dried state is detached from the outer Hadley shell save for a few isolated patches of contact. While the compressive strength of this microstructure may be high, the tensile strength and thus the modulus of rupture will be low. This strength will be related to the tensile strength of the original network of interconnected shells, with the maximum particle size, of more than 100 μm, determining the size of the largest inherent Griffith flaw. Extraneous defects in the form of the trapped air bubbles found in hand-mixed cement pastes can be removed by vacuum treatment of the fresh paste shortly after mixing. As we would expect, de-aired cement paste shows no substantial improvement in mechanical properties in the hardened state because this treatment does not alter the mechanism of hydration of the cement.

5. Conclusions

1. Rapid hydration of C_4AF in O.P.C. starts after 2–3 days and *ca.* 40 % has hydrated by the end of peak 4. Removal of CH by pozzolanic reaction accelerates and accentuates this hydration.
2. C_3S hydrates earlier and faster than most of the C_3A in O.P.C.
3. After 5 days at 20 °C and w/c 0.5, the hydration of C_3S, C_3A and C_4AF slows down, owing to the restricted access of water to the unhydrated phases.
4. C_3S, α'-C_2S, β-C_2S, C_3A and C_4AF together with their hydration products can be identified in the microstructure during the course of hydration.
5. The formation of a hydration shell after 12 h is followed by progressive dissolution of C_3S, C_4AF, and C_2S. The network of shells forms the coherent element in hydrating O.P.C. pastes and is responsible for many of their late properties.
6. The space between the shells is filled with large intergrowths of CH, ettringite rods from C_3A, AFt fibres and spherulites from C_4AF, and C–S–H from C_2S and large-grained C_3S.

7. The space within the shells is filled with inner product largely from within.

8. This inner product is detached from the shell and may consequently form an inherent Griffith flaw in hardened cement paste.

The authors wish to thank the Marine Technology Directorate of S.E.R.C. for financial support of this work, Blue Circle for the provision of cement, Mr Gutteridge for doing the quantitative X-ray diffraction analysis and identifying peak 4, figure 1, at the Cement & Concrete Association, Wexham Springs, and their colleagues for helpful discussions. In particular, they thank Karen Scrivener for commenting on the draft of the paper and Grace Halse for providing figure 1.

References

Brunauer, S. 1957 In *The science of engineering materials* (ed. J. E. Goldman), pp. 460–466. New York: Wiley.

Dalgleish, B. J., Ghose, A., Jennings, H. M. & Pratt, P. L. 1981 In *Proc. 11th Int. Conf. Sci. Ceram., Sweden*, vol. 2 (ed. R. Carlsson & S. Karlsson), pp. 297–302. Gottenburg: Swedish Ceramic Society.

Dalgleish, B. J., Ghose, A., Jennings, H. M. & Pratt, P. L. 1982*b* In *Int. Conf. Concr. at Early Ages, Paris*, vol. 1, pp. 137–143.

Dalgleish, B. J., Pratt, P. L. & Moss, R. I. 1980 *Cem. Concr. Res.* **10**, 665–676.

Dalgleish, B. J., Pratt, P. L. & Toulson, E. 1982*a* *J. Mater. Sci.* **17**, 2199–2207.

Diamond, S. 1976 In *Proc. Conf. Hydraulic Cement Pastes; their Structure and Properties, Sheffield*, pp. 2–30. Wexham Springs: Cement and Concrete Association.

Fukuhara, M., Goto, S., Asaga, K., Daimon, M. & Kondo, R. 1981 *Cem. Concr. Res.* **11**, 407–414.

Fukuhara, M., Goto, S., Asaga, K., Daimon, M., Kondo, R. & Ono, Y. 1980 *J. ceram. Soc. Japan* **88** (8), 436–440.

Ghose, A. & Pratt, P. L. 1981 In *Proc. Symp. Effects of Fly ash Incorporation in Cement & Concrete*, pp. 82–91. Boston: Materials Research Society.

Goult, D. J. & Pratt, P. L. 1983 To be presented at 1st Int. Conf. on Use of Fly ash, Silica Fume, Slag and Other Mineral By-Products in Concrete, Montebello.

Halse, Y. H. & Pratt, P. L. 1983 To be presented at 1st Int. Conf. on Use of Fly ash, Silica Fume, Slag and Other Mineral By-Products in Concrete, Montebello.

Jelenić-Bezjak, I. 1983 In *Advances in cement technology* (ed. S. N. Ghosh), pp. 397–440. Oxford: Pergamon Press.

Jennings, H. M., Dalgleish, B. J. & Pratt, P. L. 1981 *J. Am. ceram. Soc.* **64** (10), 567–572.

Jons, E. S. & Osbaeck, B. 1982 *Cem. Concr. Res.* **12**, 167–178.

Locher, F. W. 1981 In *Proc. 7th Int. Congr. Chem. Cem., Paris*, vol. IV, pp. II. 49–II. 62. Paris: Éditions Septima.

Mindess, S. & Young, J. F. 1981 In *Concrete* (ed. N. M. Newmark & W. J. Hall), New Jersey: Prentice-Hall.

Osbaeck, B. & Jons, E. S. 1980 In *Proc. 7th Int. Symp. Congr. Cem., Paris*, vol. II, pp. II. 135–II. 140. Paris: Éditions Septima.

Taplin, J. H. 1959 *Aust. J. appl. Sci.* **10**, 329–345.

Taplin, J. H. 1969 In *Proc. 5th Int. Symp. Chem. Cem., Tokyo*, vol. II, p. 249.

Discussion

J. Skalny (*Martin Marietta, Baltimore, U.S.A.*). What are the reasons for the absence of Hadley grains in pure C_3S hydration? What is the mechanistic explanation?

P. L. Pratt. In the cements we have examined, the presence of Hadley grains is associated with the formation of a network of shells around the cement particles arising from the interstitial aluminate phases. The first skeleton of this shell comes from the very early hydration of C_3A and C_4AF forming short rods of AFt within or on an amorphous gel coating. This skeleton is built on during peak 2, with middle product C–S–H from C_3S and from exposed small particles of C_2S. Finally, the hollow shells appear rapidly as C_3A hydrates during peak 3, and form rods of ettringite outside the shells, leaving space within. In pure C_3S, the absence of the aluminate phases prevents the early formation of the skeleton and also its subsequent growth. In C_3S the formation of a rim of C–S–H does slow down the hydration in the later stages, but this rim of product seems to be more flexible than the shell in O.P.C. Occasionally, hollow shells have been seen in alites and in some large particles of C_3S, but only after 14–28 days rather than 12–24 h.

P. C. Hewlett (*Cementation Research Ltd., Rickmansworth, U.K.*). Professor Pratt implied that the tensile strength of hydrated Portland cement was in some way associated with the gap between the 'outer' and 'inner' hydration layers. What is the cause of the gap and what can be done to overcome it to upgrade tensile strength?

P. L. Pratt. The cause of the gap between the outer and inner hydration layers appears to be the different mechanisms of hydration responsible for the two products. Initially, the hollow space inside the shell is associated with the onset of the rapid hydration of C_3A, to form ettringite rods, 1–2 μm long, outside the shell. The inner surface of the shell then consists of some areas of AFm, where the supply of sulphate ions is limited, and some of the less soluble remains of the cement grains. Occasionally these inner surfaces form a good basis for the nucleation and growth of the inner product inwards but most of the inner product appears to be a loose fit within the shell. Differential shrinkage and later dissolution and hydration of the less soluble remains could play a part in this separation.

If indeed this gap is associated with the low tensile strength of hydrated Portland cement, some way of overcoming it should be found. Recognized ways of increasing the strength include warm pressing the products together in an autoclave, modifying the mechanism of hydration by high shear mixing so as to break down the early AFt shells, or minimizing the amount of hydration, so that hollow shells do not form, with a polymer additive to lubricate and glue the particles together. A less drastic way would be to study the bond between the inner and outer products and, by adjusting the chemistry and mechanisms of the reactions, maximize the strength of the bond.

Phil. Trans. R. Soc. Lond. A **310**, 105–111 (1983) [105]
Printed in Great Britain

The microstructure and chemistry of tricalcium aluminate hydration

By J. E. Bailey and C. J. Hampson
Department of Metallurgy and Materials Technology, University of Surrey, Guildford, Surrey, U.K.

Electron optical techniques have been used to study the hydration of tricalcium aluminate in the presence of gypsum at various pH levels. Calcium aluminosulphate hydrates are precipitated, which have compositional variations and structures that range from highly disordered to highly ordered. Results derived from solubility measurements are able to account for the metastability of ettringite with respect to calcium aluminate monosulphate hydrate at pH more than about 12.8.

It is shown that the increased surface area of the precipitates formed at high pH is not sufficient to account for the observed increase in solubility. Consideration of the crystal structure of ettringite and the monosulphate hydrate suggests a further proposal for the type of defect occurring in the ettringite structure at high pH.

Introduction

Examination of the hydration products of ordinary Portland cement (O.P.C.) by electron microscopy has shown that the major aluminium-containing phases found during the early stages of hydration are tricalcium aluminate (C_3A), calcium aluminoferrite (C_4AF) and calcium aluminosulphate hydrate. The last is found in the form of fibres, usually radiating from another aluminium-containing phase. Analyses of these fibres indicate a composition close to that of the mineral ettringite ($Ca_6(Al(OH)_6)_2(SO_4)_3 . 26H_2O$), which has been detected in X-ray diffraction studies of the early stages of O.P.C. hydration (Chatterji & Jeffery 1963). At longer hydration times calcium aluminate monosulphate hydrate ($Ca_4(Al(OH)_6)_2SO_4 . 6H_2O$) may be detected. Bailey & Chescoe (1980) have identified both crystalline and apparently interfacial, amorphous forms of the fibrous material and their evidence that some of the amorphous fibres are hollow has been confirmed by Goodhew & Chescoe (1981).

The existence of two types of calcium aluminosulphate hydrate fibres was proposed long ago (Schwiete *et al.* 1966) and it was suggested that one type was formed by precipitation from solution and the other topochemically on the tricalcium aluminate surface. In the same paper the question of the transformation of ettringite, detected in hydrating cements by X-ray diffraction, to calcium aluminate monosulphate hydrate, in spite of the well established instability of the latter compound with respect to a mixture of ettringite and tricalcium aluminate hydrate (C_3AH_6), was raised.

We have attempted to throw some light on this situation by studying simple model systems, thus reducing the number of variables to a more manageable level. A typical Portland cement contains eight to ten elements at levels greater than 5 % by mass, combined in five or six different compounds, at least two of which exhibit several polymorphs. On reaction with water, a solution containing significant levels of up to eight ions is formed. Several of these can exist in various states depending on their concentration and environment and there are many possible interactions between them. We started our work with the simplest cement related system that would produce calcium aluminosulphate hydrates, namely the tricalcium aluminate, calcium sulphate dihydrate (gypsum), water system. From this comparatively simple system we have moved on to study the effect of pH in an attempt to produce microstructures similar to those observed in O.P.C. hydration. In order to provide some rational explanation of our results we have had to make a very careful reappraisal of the chemistry of these systems and its relation to the thermodynamic properties of the various compounds formed. Some of this work will be referred to in the discussion that follows, and details have been published elsewhere (Bailey & Hampson 1982*a*, *b*; Hampson & Bailey 1982).

Experimental

Tricalcium aluminate was prepared by firing mixtures of calcium carbonate and alumina to 1450 °C for 2 h, grinding and refiring until X-ray diffraction indicated that no other phases were present. Chemical tests for free lime (CaO) were not used as, in general, they are not sufficiently sensitive to yield useful information. For instance, at a water to solids ratio of unity, only 0.14 % of CaO in the solid phase is sufficient to saturate the aqueous phase with respect to calcium hydroxide. It was therefore our practice to make up at least one sample from each batch of tricalcium aluminate produced, in sufficient quantity for its pH to be measured. Other samples were produced in quantities *ca.* 1 g in order to allow control of the water to solids ratio and to eliminate the large random variations in CaO content that might occur if the much smaller quantities required for preparation of electron microscopy samples were used.

Samples for transmission electron microscopy (t.e.m.) were prepared by using the method of Bailey & Chescoe (1980), hydration being terminated by vacuum drying. The water to solids ratio was unity and, as explained above, each microscopy sample was a portion from a mixture containing *ca.* 1 g of solids. Details of the procedures used in preparing samples for scanning electron microscopy (s.e.m.) have already been reported elsewhere (Hampson & Bailey 1983). In this case, as only the surface of the tricalcium aluminate pellet reacts, it is not possible to control the ratio of sulphate to aluminate. The aluminate is hydrated in a solution saturated with respect to gypsum, for a given time. In order to simulate the effect of a limited supply of gypsum, such as exists in Portland cement, the pellet of aluminate could be moved at a given time from a solution saturated with respect to gypsum to one of lower sulphate concentration. This has not yet been attempted as it introduces two further variables into the system, namely a second hydration time and the sulphate activity of the second solution.

Results

The results of the electron microscope examinations have already been outlined in the literature (Hampson & Bailey 1983) but it is instructive to assemble some of them in the form of a diagram. In figure 1 the t.e.m. observations concerning the hydration of tricalcium aluminate/gypsum mixtures are summarized as a function of sulphate to aluminate ratio and hydration time. Hexagonal platelet phases are only observed in the region below the dashed line. At sulphate to aluminate ratios greater than unity the only phases observed are calcium aluminosulphate hydrate fibres and gypsum. The absence of tricalcium aluminate has been confirmed by X-ray and infrared examinations (Hampson & Bailey 1983). The initial pH of all these systems is 11.5. The calcium aluminosulphate hydrate fibres were generally 1 μm long × 0.1 μm diameter. Microanalysis showed that the platelet phases contained sulphur.

When the pH of the system was raised by the addition of sodium or calcium hydroxide, or both, the picture changed. In the pH range 12.4–13.0 smaller fibres of calcium aluminosulphate were observed, growing on the surface of the tricalcium aluminate grains. The rate of consumption of the tricalcium aluminate was greatly reduced compared to that at pH 11.5 and the addition of the gypsum at the higher pH range could then be said to retard the tricalcium aluminate hydration. The reduction in fibre size to 0.2 μm × 0.05 μm was also noted by Schwiete *et al.* (1966) and Mehta (1973). There was evidence of a precursor platelet phase being formed at the surface of the tricalcium aluminate grains, similar to that observed by Tinnea & Young (1977) and, during the hydration of tricalcium aluminate without gypsum, by Breval (1976).

At pH 13 the tricalcium aluminate was consumed rapidly and no fibres or platelets were observed on its surface.

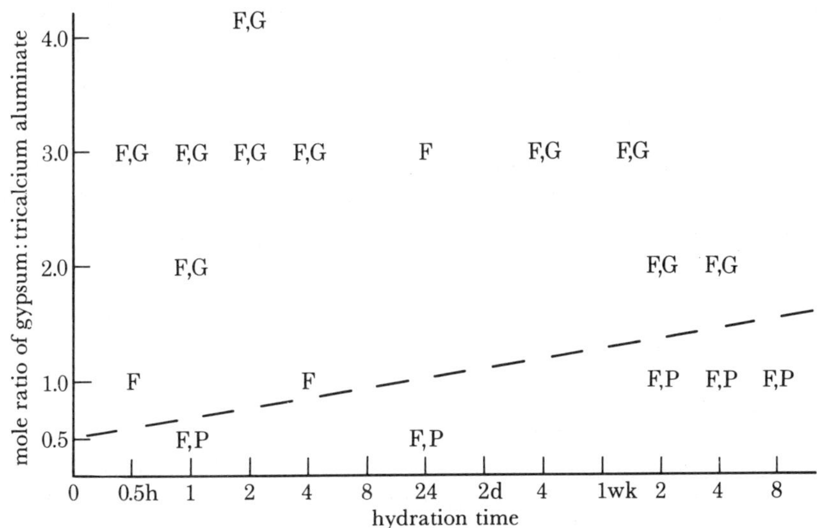

FIGURE 1. Phases observed during the hydration of mixtures of tricalcium aluminate and gypsum. Abbreviations: F, fibres; G, gypsum; P, platelets.

DISCUSSION

These results, and a careful reassessment of the chemical data available for the aqueous phase of the system $CaO/Al_2O_3/SO_3/H_2O$ have led us to make several proposals as a basis for a model of the hydration process of tricalcium aluminate in the presence of gypsum (Hampson & Bailey 1983). These are:

(i) the first hydration product is a platelet phase, being a solid solution between

$$Ca_2(Al(OH)_6)_2 . 2H_2O \quad \text{and} \quad Ca_4(Al(OH)_6)_2 . (OH)_2 . 12H_2O$$

(or, in the usual cement chemists notation: C_2AH_8 and C_4AH_{19}).

(ii) The location of the next product depends on the solubility of this phase, which in turn depends on the pH of the surrounding aqueous phase. In particular, it depends on the aluminate ion solubility (Bailey & Hampson 1982*b*).

(iii) At low pH (*ca.* 11.5) this second product is produced by through solution precipitation. It consists of crystalline fibres of calcium aluminosulphate hydrate with a crystal structure similar to ettringite, but with a composition deficient in both calcium and sulphate ions with respect to that mineral. The limit of this non-stoichiometry appears to be a composition of approximately $Ca_4(Al(OH)_6)_2SO_4.xH_2O$, i.e. a material with the same sulphate to aluminate ratio as calcium aluminate monosulphate hydrate.

(iv) At pH *ca.* 12.5 the second product is formed at the tricalcium aluminate surface and covers or replaces the layer of platelets. The fibres are smaller than those observed at lower pH and have a less ordered structure. The latter is reflected in the increased solubility of such materials, and their decreased stability in the electron microscope. We have previously suggested that this disorder is due to a deficiency of aluminium in the fibres (Hampson & Bailey 1982) and noted an empirical correlation between the solubility product of the ettringite-like phase and the product of the activities of the sulphate and hydroxide ions in the surrounding solution.

The first three proposals appear to correspond well with our observations and are based on accepted theories of physical chemistry. The fourth, however, raises certain problems. The proposed aluminium deficiency in the fibres formed at high pH was made on the basis of the reduction in aluminate activity observed in solutions in equilibrium with ettringite as the pH is raised from *ca.* 11.5 to 12.5. However, it is observed that it is in alkali-containing systems with pH *ca.* 12.5 that the greatest increase in the ettringite solubility product occurs. Under these conditions the aluminate level in solution rises with increased pH (Bailey & Hampson 1982*b*). Such a rise in aluminate activity in solution does not seem consistent with a reduction in the aluminium content of the fibres. In general the composition of non-stoichiometric precipitates reflects that of the solution around them; such an effect is, for instance, observed in apatite precipitates (McClean & Budy 1964). Thus while we retain the hypothesis that the increase in solubility of ettringite at high pH, and the other changes observed, are due to the observed disorder in its crystal structure, we wish to revise our opinion as to the nature of these defects. We will therefore devote the remainder of this paper to a careful consideration of some of the observations relating to calcium aluminosulphate hydrates at high pH.

The most readily quantifiable parameter is the increase in molar free energy of the solid phase implied by its increased solubility product (K_s). The latter varies from 10^{-45} to 10^{-35} as the pH is raised from 11.5 to *ca.* 13 (Hampson & Bailey 1982), which corresponds to a molar free energy increase ΔG, from -250 to -170 kJ/mol of calcium aluminosulphate hydrate, when formed from the appropriate ions at unit activity. (The relation $\Delta G = RT \ln K_s$ is derived in the majority of physical chemistry texts, for example, Moore (1972)). At pH *ca.* 12.8 the fibrous calcium aluminosulphate hydrate is unstable with respect to a mixture of the plate like monosulphate and gypsum (Jones 1944; Kalousek 1941). The change in molar free energy when the fibrous 'ettringite' phase is transformed to such a mixture can be calculated from the molar free energies of formation of the species involved, as in table 1. The desired result is the quantity ΔG_4, which is clearly equal to $\Delta G_3 - \Delta G_2$. ΔG_2 is the quantity already obtained from the solubility products of the fibrous aluminosulphate hydrate phase. To obtain ΔG_3 we begin by applying our calculation of solubility product to the data of D'Ans & Eick (1954) for solutions in equilibrium with calcium aluminate monosulphate hydrate. An average value of 10^{-27} is obtained for the solubility product of the monosulphate, corresponding to a free energy change of -154 kJ/mol. To this must be added twice the molar free energy change for the formation of gypsum from ions at unit activity, namely -52 kJ/mol, giving a total of -206 kJ/mol for ΔG_3. Thus if $\Delta G_2 = -250$ kJ/mol then $\Delta G_4 = -206-(-250) = +44$ kJ/mol whereas if $\Delta G_2 = -170$ kJ/mol $\Delta G = -36$ kJ/mol. In the first case the 'ettringite' is the stable phase, while in the second the mixture of monosulphate and gypsum has the lower free energy. The change from one situation to the other occurs when $\Delta G_2 = -206$ kJ/mol. This corresponds to K_s *ca.* 10^{-36}, a value that is observed at pH 12.5 in systems saturated with gypsum. Thus the observed metastability of calcium aluminosulphate fibres at pH more than about 12.8 may be explained by the increased molar free energy of such materials.

The second characteristic of fibrous calcium aluminosulphate hydrate precipitates is the reduction in size that is observed as the pH at which they are formed is raised. The increase in the vapour pressure of liquid droplets and in the solubility of precipitates as their particle size

Table 1. Free energy changes in the formation of ettringite and calcium aluminate monosulphate hydrate

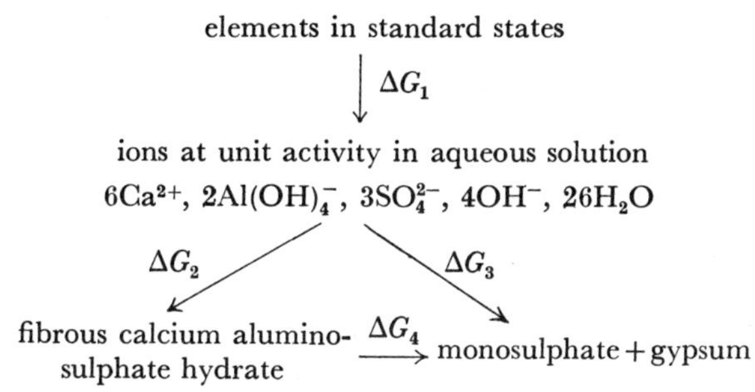

is decreased is well known and it has been suggested that the change in solubility product that we have observed for ettringite is due to the change in fibre size. From an estimate of the surface energy we can calculate the increase in the energy of the fibres due to their decreased size. A cylindrical fibre of radius r and length l, with surface energies γ_e, and γ_s for the ends and sides respectively has a minimum surface energy when

$$\gamma_e/\gamma_s = l/2r,$$

and in this state the total surface energy of the ends is half that of the sides. For ettringite the easier of the two surface energies to estimate is that of the cylindrical surface, γ_s. Analysis of the crystal structure of ettringite (Moore & Taylor 1970) shows it consists of positively charged calcium aluminate hydrate columns, joined laterally by hydrogen bonds and separated by sulphate ions. These columns are parallel to the axis of the fibres and it is therefore the hydrogen bonds that have to be broken if the cylindrical surface of the fibres is to be created. Each unit cell of the structure consists of one column, approximately a cylinder of radius 0.56 nm, length 2.15 nm, containing 52 water molecules in addition to the calcium and aluminate ions plus its associated sulphate ions. The cylindrical surface of each cylinder amounts to 8×10^{-18} m^2 and is crossed by 52 hydrogen bonds, each with an energy 0.1 eV or 1.6×10^{-20} J (Kittel 1971). To a first approximation the surface energy γ_s is therefore

$$\gamma_s = \left(\frac{52}{2}\right)\frac{1.6 \times 10^{-20}}{8 \times 10^{-18}} = 0.05\,\mathrm{J\,m^{-2}}.$$

Ignoring the ends, the specific surface of a cylindrical fibre of radius r and density ρ is equal to $2/r\rho$. For ettringite $\rho = 1764$ kg m^{-3} and r is observed to vary from 0.05 µm to 0.2 µm. The corresponding surface areas are 23 and 57 m^2/g respectively. The relative molecular mass of ettringite is 1254 so the surface area of 1 mole of the fibres changes by 34×1254 m^2 as the pH of formation is raised from 11.5 to 12.5. If the surface energy is 0.05 J m^{-2} the corresponding increase in energy is 2.1 kJ/mol. Over the same pH range the observed increase in molar free energy, calculated from the solubility product, is 60 kJ/mol so it is not possible that the decrease in fibre size accounts for the increase in solubility.

We have already suggested (Hampson & Bailey 1982) that at high pH levels there is some form of defect in the calcium aluminate hydrate columns that form the basis of the ettringite structure and that it is this defect that increases the free energy. Since the monosulphate is stable at high pH, its structure is presumably not affected in the same way. Studies of the related compound $Ca_4(Al(OH)_4)_2(OH)_6.6H_2O$ (C_4AH_{13}) (Ahmed & Taylor 1967) showed it to consist of positively charged sheets of formula $(Ca_2Al(OH)_6^+)_n$ interleaved with negative ions (OH^-, SO_4^{2-} or $Al(OH)_4^-$) and water molecules. In the case of ettringite the basic unit of the crystal structure is, as already described, a column of formula $(Ca_3Al(OH)_6^{3+})_n$. Both these columns and the sheets in the monosulphate structure contain water. Aluminate ions in the aqueous phase of Portland cements and related systems exist in the form $Al(OH)_4^-$, while a significant fraction of the calcium in solution is in the form of the ion pairs $CaOH^+$ and $CaSO_4$ (Bailey & Hampson 1982*a*). This fraction increases as the pH and sulphate activity of the aqueous phase increase.

In order to form sheets, as in the monosulphate, each aluminate ion must combine with two Ca^{2+} and two OH^- ions or two $CaOH^+$ ion pairs. To form columns, as in ettringite, however, three Ca^{2+} ions and two OH^- ions are required, or two $CaOH^+$ ion pairs and one Ca^{2+} ion. It seems likely that it is the reduction in Ca^{2+} ion activity at elevated pH and sulphate levels that leads to the disorder in the fibrous calcium aluminosulphate hydrates, and that the defect is again a calcium deficiency in the columns, accompanied presumably by a deficiency of sulphate ions to maintain the charge balance. In this case it is caused by a lowering of the Ca^{2+} activity in the aqueous phase, whereas at lower pH the cause may be a kinetic one relating to the comparatively rapid dissolution of the tricalcium aluminate.

Conclusions

The observed increase in solubility product, and hence molar free energy, of fibrous calcium aluminosulphate hydrates at high pH (*ca.* 12.8) is sufficient to account for their metastability with respect to calcium aluminate monosulphate hydrate and gypsum. It is unlikely that the increase in molar free energy is due to the apparently increased surface area of the fibres formed at high pH.

Consideration of the crystal structures of ettringite and the monosulphate, and of the ions present in solution, leads us to suggest that disorder in the structure of fibres formed at high pH is in the form of calcium deficiency in the calcium aluminate hydrate columns and is due to the reduction of Ca^{2+} activity in solution. This is the same type of defect as was previously proposed for the initial fibrous precipitate formed in tricalcium aluminate/gypsum mixtures at pH *ca.* 11.5.

References

Ahmed, S. J. & Taylor, H. F. W. 1967 *Nature, Lond.* **215**, 622.
D'Ans, J. & Eick, H. 1954 *Zem.-Kalk-Gips* **7**, 449.
Bailey, J. E. & Chescoe, D. 1979 *Proc. Br. Ceram. Soc.* **28**, 165.
Bailey, J. E. & Hampson, C. J. 1982*a* *Cem. Concr. Res.* **12**, 227.
Bailey, J. E. & Hampson, C. J. 1982*b* *Int. Semin. on Calcium Aluminates*. Politechnico di Torino, September 1982, Torino, Italy.
Breval, E. 1976 *Cem. Concr. Res.* **6**, 129.

Chatterji, S. & Jeffery, J. W. 1963 *J. Am. ceram. Soc.* **46**, 268.
Goodhew, P. J. & Chescoe, D. 1982 *Inst. Phys. Conf. Ser.* **61**, 351.
Hampson, C. J. & Bailey, J. E. 1982 *J. Mater. Sci.* **17**, 3341.
Hampson, C. J. & Bailey, J. E. 1983 *J. Mater. Sci.* **18**, 402.
Jones, F. E. 1944 *J. phys. Chem.* **48**, 311.
Kalousek, G. L. 1941 Ph.D. thesis, University of Maryland, U.S.A.
Kaye, G. W. C. & Laby, T. 1968 *Tables of physical and chemical constants*, 13th edn, pp. 175–184. London: Longman.
Kittel, C. 1971 *Introduction to solid state physics*, p. 126. New York: John Wiley.
McClean, F. C. & Budy, A. M. 1964 *Radiation, isotopes and bone*, p. 64. New York: Academic Press.
Mehta, P. K. 1973 *J. Am. ceram. Soc.* **56**, 315.
Mehta, P. K. 1976 *Cem. Concr. Res.* **6**, 169.
Moore, A. E. & Taylor, H. F. W. 1970 *Acta crystallogr.* B **26**, 386.
Moore, W. J. 1972 *Physical Chemistry*, 5th edn, p. 310. London: Longman.
Schwiete, H. E., Ludwig, U. & Jager, P. 1966 *Highway Res. Board. Special Report* no. 90, p. 353, U.S.A.
Tinnea, J. & Young, J. F. 1977 *J. Am. ceram. Soc.* **60**, 387.

Discussion

J. Bensted (*Blue Circle Technical Research Division, Greenhithe, U.K.*) Perhaps, in view of the variability in composition found in the ettringite phase, a change in name away from 'ettringite' would be useful?

J. E. Bailey and C. J. Hampson. We agree; these calcium aluminosulphate hydrates should not be referred to as ettringite.

Phil. Trans. R. Soc. Lond. A **310**, 113–125 (1983) [113]
Printed in Great Britain

Strength and toughness in ceramic systems

By R. W. Davidge
Materials Development Division, Building 552, A.E.R.E. Harwell, Didcot, Oxon, OX11 0RA, U.K.

There has been a steady development and use of fracture mechanics, for describing the strength of brittle materials, ever since its inception by Griffith (1920). Materials considered include glass, ceramics, cement, brittle polymers, brittle metals, and a wide range of composite materials such as fibrous composites. The complexity of the fracture process generally increases as one passes along this list. A basic concept of fracture mechanics is a fracture surface energy or fracture toughness. Ideally this is a material constant but this is rarely true in practice. This paper summarizes the current state of understanding for ceramics and discusses a number of specific problem areas. This information should thus act as a signal to workers using fracture mechanics on cementitious materials. Problems considered include environmental sensitivity, statistical variations in strength, crack length effects, the concept of a flaw for porous materials, plastic effects, and transformation toughening.

1. Introduction

Fracture mechanics provides the basic framework for understanding the strength and toughness of ceramics. The tensile fracture strength of a ceramic σ_f can be understood in terms of the Griffith equation

$$\sigma_f = \frac{1}{Y}\left(\frac{2E\gamma_i}{C}\right)^{\frac{1}{2}} = \frac{K_{Ic}}{YC^{\frac{1}{2}}}, \tag{1}$$

where Y is a geometrical constant, E is the Young modulus, γ_i is an effective surface energy, C is a flaw size, and K_{Ic} is a stress intensity factor or fracture toughness. This equation implies that the surface energy or fracture toughness of the material is a constant, so that provided that the flaw size is known, the fracture strength can be estimated.

In the original experiment, Griffith (1920) produced thin slit-like cracks throughout the thickness of glass spheres and cylinders. The specimens were broken by the application of an internal pressure, with a compressive axial stress being applied as an additional variable for the cylindrical specimens. Griffith observed that the product $\sigma_f C^{\frac{1}{2}}$ was essentially constant in his experiments and furthermore that the effect of a compressive stress in the cylindrical specimens did not affect this observation. Griffith, in fact, tested only ten specimens and it is therefore, perhaps, not surprising that he did not discover some of the complicating factors that are significant in understanding the strength of ceramic materials under general conditions. He did, however, obviously recognize the importance of a number of key variables and, apart from the state of stress, he also mentioned the effects of the range of crack sizes used, the time taken to fracture the specimens, and the test environment.

Complicating factors affecting the strength of ceramics are manifested in two main ways. First, strength is generally time dependent in that the application of a tensile stress on a component causes a gradual diminution of its capability to withstand further stress without rupture. Second, there is a relatively large statistical variation in the strength of a batch of otherwise identical specimens. These factors can both be considered with reasonable success at an empirical level so that good prediction of mechanical performance may be made. Additionally there is an increasing awareness of the importance of various microstructural features and how these affect fracture. This paper is concerned with a review of some of the more important recent developments in understanding in these areas.

2. Subcritical crack growth

When a brittle ceramic specimen, with a large preformed crack, is fractured in tension, in vacuum or inert atmosphere, its behaviour follows classical expectations of the Griffith equation (1). The fracture surface energy is essentially constant, and specimens with similar-sized cracks fail at a similar force.

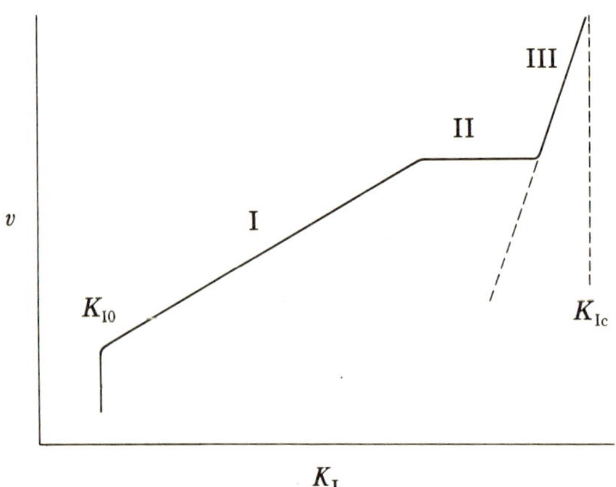

Figure 1. Idealized stress intensity factor–crack velocity (K/v) diagram for ceramics (logarithmic scales).

Under more aggressive conditions, for example wet environments for oxides, or at high temperatures for all ceramics, the situation is very different. Data relating crack velocity to stress intensity factor have been obtained for a wide range of ceramic materials by using standard fracture mechanics tests, particularly the double torsion test and the double cantilever beam test. The classical behaviour observed is shown in figure 1, where there are three well defined regions. Above a threshold stress intensity factor the crack velocity in region I is proportional to the nth power of the stress intensity factor. In region II the crack velocity is independent of stress intensity factor, and in region III the behaviour is similar to that in region I but with a significantly higher slope. Fracture finally occurs at the critical stress intensity factor K_{Ic}. The rate of crack growth in region I is reaction rate controlled, whereas in region II it depends on the diffusion of a corrosive species to the crack tip. In most situations the corrosive agent is water and region II occurs at higher crack velocities with increase in water content. Region III represents an inherent behaviour of the material and is the only stage observed when testing is done in vacuum.

The important effects relating to time dependence of strength are generally concerned almost solely with region I in that the behaviour associated with other regions occurs during very short times. This has led to the emergence of the parameter n as all-important in determining the time dependent behaviour of ceramics.

For simple systems such as glass, where the effects of microstructure are eliminated, there is very good agreement between basic theory and experimental observation (Wiederhorn 1980). The theory is based on a stress enhanced chemical reaction between water and the highly stressed material near the crack tip. The basic rate equation from chemical reaction rate theory can be modified to reflect physical and chemical processes that occur at the crack tip. The theory is consistent with data for crack growth dependence on temperature, applied stress intensity factor, and the concentration of water in the environment.

3. Statistical variations in strength

The flaws from which fracture originates in normal engineering ceramics are small (typically 10–100 μm) and numerous. Fracture initiates from the largest effective flaw and therefore it is necessary to consider strength on a statistical basis. With reference to equation (1) statistical variations in strength can be discussed in terms of K_{Ic} and C (neglecting any errors of physical measurement, which are small). The determination of K_{Ic} from standard fracture mechanics tests requires the introduction of a large crack into the specimen that is much larger than the inherent flaw size. Data on the statistical variation of the critical stress intensity factor has been measured by Pankow & Finnie (1979) for polycrystalline alumina. Their results are shown in figure 2. The K_{Ic} data were generated from specimens with sharpened notches (radius 20 μm) and are compared with strength data for smooth beams broken in three-point bending. The scatter in K_{Ic} is four times less than the scatter in strength. This suggests that for large sharp cracks K_{Ic} may be essentially constant. This is intuitively reasonable because the crack is sampling a statistically large area of the specimen as it propagates. Therefore the scatter in strength must be due mainly to the variation, between specimens, in the largest flaw size.

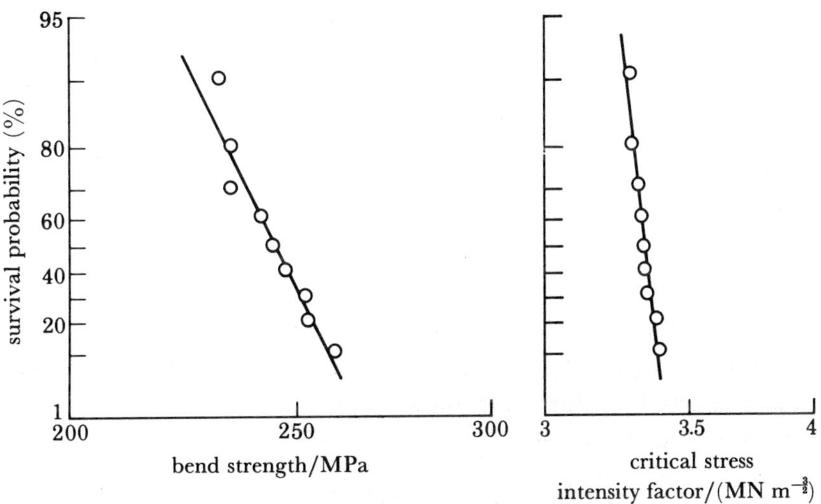

FIGURE 2. Statistical variations in strength and stress intensity factor for alumina (Pankow & Finnie 1979).

The statistical treatments used to describe the strength of ceramics are based on weakest link theory and on assumptions referred to in statistical theory as the Poisson postulates. The general theory of statistics of extremes is given in the book by Gumbel (1958) and the weakest link theory of strength, pioneered by Weibull (1951), is one aspect of this. In essence the problem is:

given a particular distribution of property values (for example strength), what is the minimum value expected for a particular number of samples? For example, one might be required to estimate the highest operating stress that could be applied to a component to give a failure rate of less than 1 in 10^6 from a number of observations of strength (typically, less than 100).

The theory assumes that the material can be divided into a number of small regions (which could be volume $\mathrm{d}V$ or area $\mathrm{d}A$ depending on whether volume or surface flaws are controlling) that act independently, each containing a flaw associated with a particular strength S. The strength of the whole is then determined by the strength of the weakest region, as in the links of a chain. The theory as applied to ceramics considers a function $g(S)$ such that the number of flaws per unit volume (or area) associated with a strength between S and $S+\mathrm{d}S$ is $g(S)\,\mathrm{d}S$. If $\mathrm{d}\varphi$ is the probability of failure of an element

$$\mathrm{d}\varphi(S) = \mathrm{d}V \int_0^S g(S)\,\mathrm{d}S, \tag{2}$$

with a survival probability $1-\mathrm{d}\varphi(S)$. In material of volume V there are $V/\mathrm{d}V$ volumes $\mathrm{d}V$. The probability $(1-\varphi)$ that no volume has a strength lower than S is the product of the separate probabilities, and

$$1-\varphi = (1-\mathrm{d}\varphi)^{V/\mathrm{d}V} = \left\{1-\mathrm{d}V \int_0^S g(S)\,\mathrm{d}S\right\}^{V/\mathrm{d}V}. \tag{3}$$

This reduces to, as $V/\mathrm{d}V \to \infty$,

$$1-\varphi = \exp\left\{-V \int_0^S g(S)\,\mathrm{d}S\right\}. \tag{4}$$

The theory of the statistics of extremes considers three kinds of asymptotic function for $g(S)$: the first two refer to variates that are unlimited in extreme values; for distributions that are bounded at one extreme the so-called asymptotic function of the third kind holds. So, this third function is of most relevance to the strength of engineering ceramics, which must have a tensile strength not less than 0. Analysis shows that the simplest function for $g(S)$, relevant to a lower limit S_u, is given by

$$\int_0^S g(S) = \left(\frac{S-S_u}{S_0}\right)^m, \tag{5}$$

where S_0 is a 'scale' parameter and m the 'shape' parameter. This will be recognized as the well-known Weibull distribution; m is usually known as the Weibull modulus and increases with decreasing variability in strength.

Weibull statistics are widely applied to the strength of ceramics, and for many data there is good agreement between experimental results and the above function. Often, the lower limit of strength S_u is taken as zero and thus, rearranging equations (4) and (5) gives a proportionality between survival probability $P_s = 1-\varphi$ as

$$\ln \ln (1/P_s) \propto m \ln S. \tag{6}$$

Typical data fitting this relation are shown in figure 3. The prediction of stress levels for high survival probabilities is possible but this should be treated with some caution because extrapolations way beyond the experimental data are involved and there is little evidence to indicate that the predictions are accurate.

An additional attraction of Weibull statistics is that, for materials that are consistent with the theory, it is possible to predict effects on strength of the state of stress (for example, uniaxial, biaxial), the distribution of stress, and the specimen size. Here, however, we are concerned more with the average rather than extreme values of strength. Both these effects are a manifestation of the fact that the greater the volume of material under stress, or the greater the number of flaws perpendicular to the tensile stress, the higher the chance of finding a large flaw.

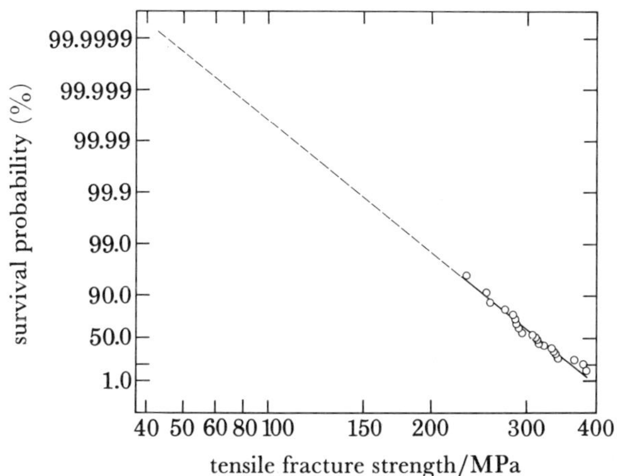

FIGURE 3. Statistical data for the tensile strength of REFEL silicon carbide according to equation (6) (Davidge 1979).

4. DESIGN PROCEDURES FOR CERAMICS

Although the experimental data that can be obtained on the time dependence of strength and the statistical variations in strength are based partly on empirical relations, they can be used for engineering design purposes (Davidge 1979).

Useful design procedures are now developed such as the strength–probability–time (s.p.t.) diagram (Davidge *et al.* 1973) and the proof stress diagram (Wiederhorn 1974), and these are being used increasingly for the routine evaluation of engineering ceramic components. Examples are shown in figures 4 and 5 for a vitreous bonded alumina ceramic as used for grinding wheels The s.p.t. diagram shows that under use at the maximum recommended operating speeds (with the maximum stress one quarter of the mean bursting strength of the wheel) the probability of survival is extremely high. With the additional safeguard of a proof test (which is common practice in the industry), the proof test diagram indicates what level of proof test ratio (proof stress: maximum operating stress) is required to guarantee a specific minimum lifetime for the grinding wheel. This is only one specific example of an engineering approach to the use of ceramics and there are many others.

5. MICROSTRUCTURAL EFFECTS

In this section we mention a number of microstructural effects that are relevant to the understanding, at a more fundamental level, of the mechanical properties of ceramics.

(*a*) *Changes in fracture toughness with crack length*

Data for the critical stress intensity factor are obtained generally from specimens with large preformed cracks, but there are complications at flaw sizes comparable in size to microstructural features. The surface energy for a grain size crack should approximate to the cleavage or grain boundary surface energy (γ_0), which is usually at least an order of magnitude less than the effective surface energy for macroscopic fracture (γ_1). Further propagation of the crack beyond a grain size dimension has a greater surface energy requirement and this may be a simple step function, or more complicated as indicated in figure 6, where it is envisaged that γ rises to γ_1

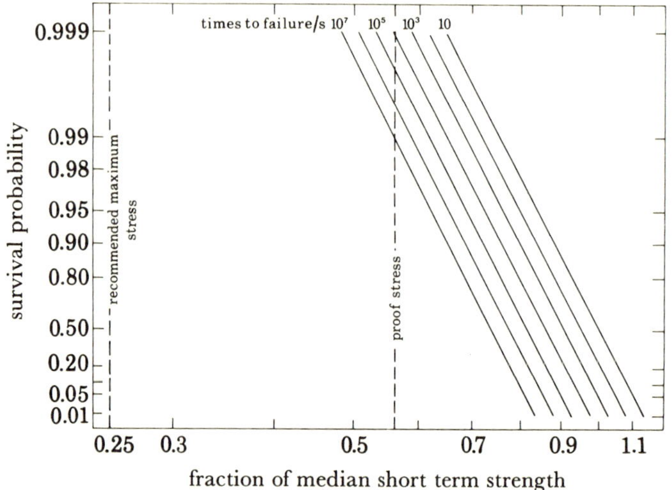

FIGURE 4. Strength–probability–time diagram for vitreous bonded alumina ceramic (McLaren *et al.* 1978).

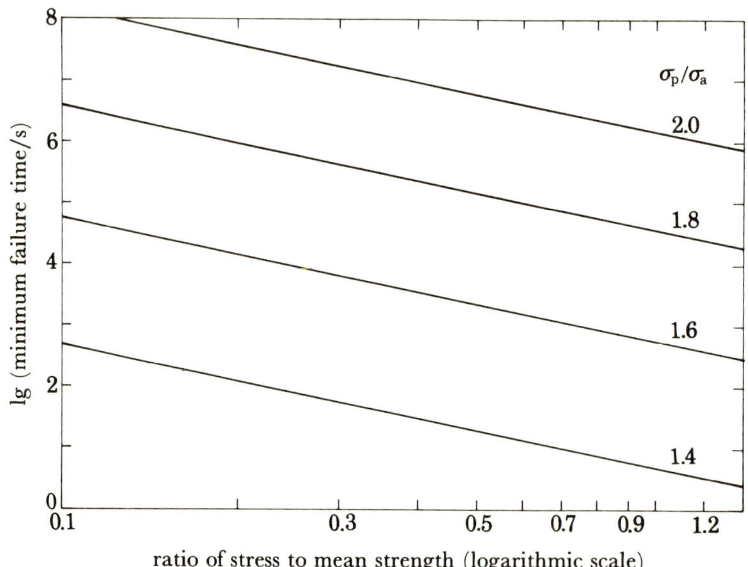

FIGURE 5. Proof test failure diagram for vitreous bonded alumina.

over a length of a few grain dimensions. Strength is controlled by the maximum value of γ/C and in figure 6 the value at points P, Q and R, for the three curves drawn. Fracture may thus be preceded by a period of subcritical crack growth based on microstructural considerations. (This is in addition to the environmental stress corrosion effects discussed above.) Thus for inherent flaws it is conceptually difficult to discuss strength in terms of an independent crack size and stress intensity factor in that these two parameters interact strongly.

A serious consequence of this effect relates to the different engineering design behaviour of large cracks, as used to generate engineering design data, and small cracks from which the fracture originates in normal specimens. Values of n obtained from K/v diagrams should thus be treated with some caution, and it is often preferable to use data for n obtained from strain rate dependence of strength or delayed fracture measurements, which relate to the behaviour of the inherent flaws. For example, in alumina it has been found that the value of n obtained

P
R
Q
γi
γ0
surface energy
0
2
4
6
N
crack length (grain diameters)

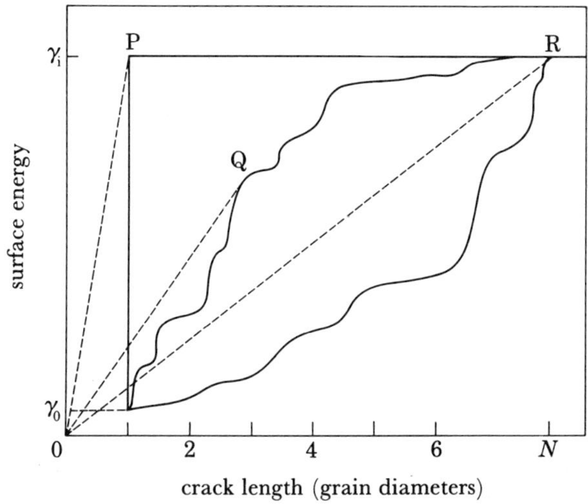

FIGURE 6. Variation of effective surface energy with crack length.

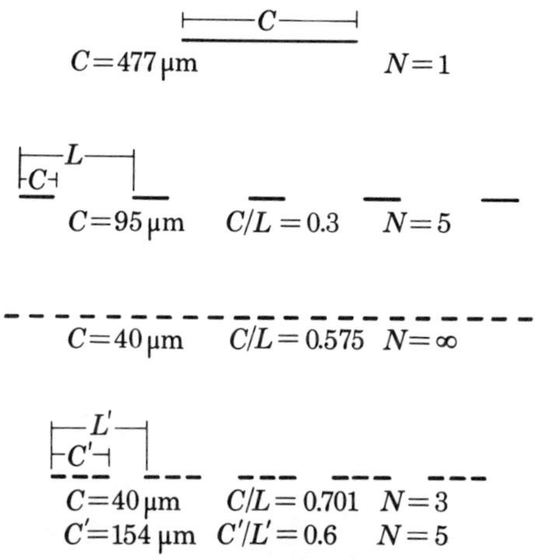

C
C=477 μm
N=1
L
C
C=95 μm
C/L=0.3
N=5
C=40 μm
C/L=0.575
N=∞
L'
C'
C=40 μm
C/L=0.701
N=3
C'=154 μm
C'/L'=0.6
N=5

FIGURE 7. Various linear arrays of microcracks that would all propagate at the same stress (Okada & Sines 1983).

from strain rate variations is approximately one half that obtained from double torsion tests (Davidge *et al.* 1973). The K/v data from macroscopic cracks would thus lead to a highly optimistic performance of components.

(*b*) *Effects of crack density*

A further complication arises from the density of inherent cracks. The theory assumes that these flaws do not mutually interact. However, when the crack spacing is similar to the crack size this is not true. Okada & Sines (1983) have identified a relatively high density of active microcracks in polycrystalline alumina and a number of close but small microcracks can coalesce and lead to failure. Figure 7 shows three such arrays of cracks that could all propagate in a short time at the same stress as the single large crack at the top of the figure. Note that this has serious consequences for non-destructive testing techniques. The nil-observation of flaws above a particular size in a sample could lead to an optimistic prediction for strength if fracture was initiated from closely-spaced smaller flaws. An additional problem is that the times to failure under delayed fracture conditions for the various crack arrays vary by an order of magnitude.

Related effects are expected to be significant in porous materials. For porosity values larger than 5–10 %, the porosity in ceramics is of the open, connected type. In highly porous materials the pores, which act as flaws, are all interconnected and the entire specimen can be regarded as being permeated by a single large flaw.

(*c*) *Zirconia toughening of ceramics*

The presence of cracks in ceramics is not always detrimental to their mechanical properties and, for zirconia toughening, the presence of a high density of very fine cracks can be positively beneficial.

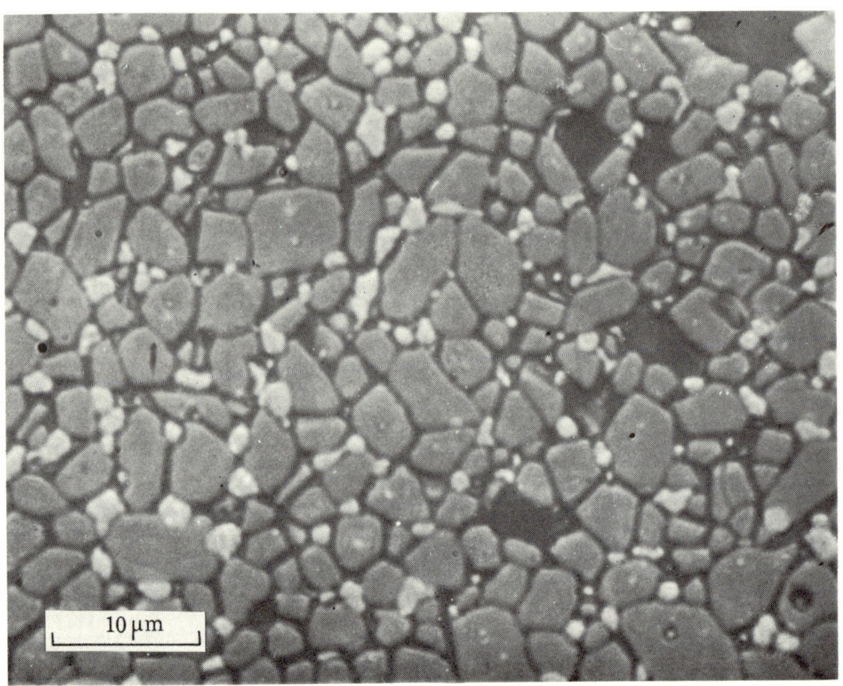

Figure 8. Optical micrograph of alumina toughened with 20 % tetragonal zirconia.

Zirconia exists in several crystallographic forms. Normally, amounts of stabilizing oxides such as MgO, CaO or Y_2O_3 are incorporated to stabilize the cubic phase. Pure zirconia, however, exists in a tetragonal crystallographic form at temperatures above 1000 °C but in a monoclinic form at lower temperatures. The transformation, from one phase to the other on cooling, which is of a martensitic type, produces large volume increases (*ca.* 5 %) and high shear strains. When tetragonal zirconia is incorporated as a second phase into a host matrix, two different effects can occur depending upon whether the zirconia is retained as the tetragonal phase or has transformed to the monoclinic phase in the ceramic as fabricated. In the first case one can get toughening plus strengthening, in the second one can find toughening but commonly without strengthening.

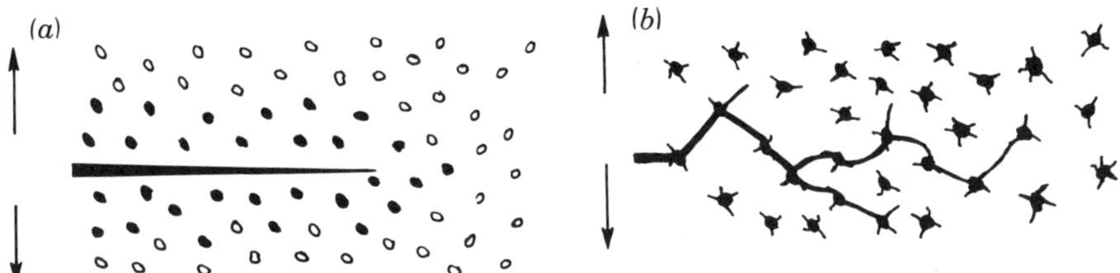

FIGURE 9. Toughening mechanisms for zirconia toughened ceramics. (*a*) Tetragonal zirconia particles (open areas) transform to monoclinic zirconia (black areas) in the vicinity of the crack tip. (*b*) Monoclinic zirconia particles with associated microcracks lead to deflexion and bifurcation of the main crack.

The essential feature of the toughening plus strengthening mechanism is that the zirconia particle size must be kept very fine, typically 1 μm or less. The elastic constraint of the matrix material may be great enough to suppress the phase transformation when the material is cooled to ambient temperatures after fabrication. In material with the optimum microstructure, figure 8, the transformation is triggered by the application of stress. In regions near the tips of the cracks, the magnified tensile stresses can reduce the constraint of the matrix on the particle so that the transformation occurs. The result is that a partly compensating compressive stress is set up just ahead of the crack tip, which makes crack propagation much more difficult, thus increasing both toughness and strength. Normally the transformed particles are restricted to a zone *ca.* 10 μm wide adjacent to the crack face. Toughness and strength are increased by about a factor of 2 with absolute values depending on the inherent properties of the matrix material. The basis of this mechanism is sketched in figure 9.

For the straightforward toughening (but not strengthening) mechanism the zirconia particles have already transformed in the material in its state as fired. The transformation and volume increase are associated with the generation of a network of localized cracks radiating from the transformed particles. The essential feature here is that any major crack is continually deflected and bifurcated on approaching the toughening particles, which leads to the increase in toughness, as indicated in figure 9. This mechanism is particularly significant for bestowing dramatic improvements in thermal shock resistance.

(*d*) *Effects of microstructural and chemical changes at high temperature*

In the discussion so far the only changes occurring in the material involve the subcritical growth of microcracks up to the point of catastrophic failure. Whereas this situation is obtained for low-to-moderate temperatures, it is much more complicated at higher temperatures where additional effects can occur involving microstructural or chemical changes to the material. There are a number of important effects including:

(*a*) localized plastic flow, which can induce crack healing or the generation of creep induced voids;

(*b*) chemical changes, particularly oxidation, which again can produce crack healing but also the generation of new defects;

(*c*) microstructural changes such as grain growth, phase changes or surface evaporation.

Any of these changes may be either deleterious or advantageous with respect to strength.

The important consequence is that the original flaw population present at ambient temperature cannot necessarily be regarded as relevant in controlling the mechanical response of a material to stress at high temperatures. This means that the valuable techniques mentioned above such as proof testing or non-destructive testing have limited validity for high temperature performance prediction. This is because the ranking of the strength of a set of components is different at low temperature compared with high temperature. A dramatic example of this for hot pressed silicon nitride is shown in figure 10. Data for two sets of specimens are compared at 25 °C and 1200 °C, before and after proof testing at 25 °C. In the data at 25 °C the strength of the proof-tested specimens is highly truncated, and in agreement with theory, indicating the value of the proof test. On the other hand, similar data for specimens tested at 1200 °C, which had first been heated for 30 min at 1200 °C, show virtually identical strength distributions irrespective of proof-testing. This indicates strongly that the original flaws are not relevant in controlling strength under high temperature conditions.

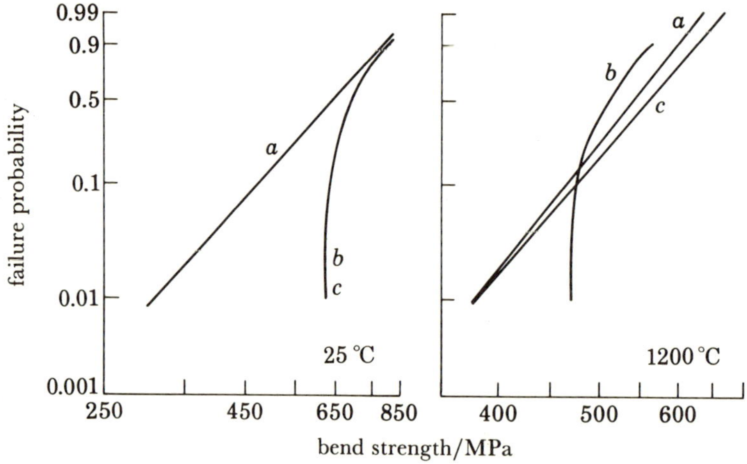

FIGURE 10. Effects of proof testing at 25 °C on the strength of silicon nitride tested at 25 °C, and after 30 min at 1200 °C. (*a*) Samples not proof tested. (*b*) Theoretically predicted curve after proof testing. (*c*) Proof tested samples.

One thus has a daunting list of possibilities for potential high temperature effects that control strength and it is not surprising that performance prediction techniques under these conditions are still in their infancy. This is however clearly a priority area for research if the reliability of structural components for high temperature applications is to be derived with any confidence.

6. Relevance to cement

Cement is traditionally regarded as a brittle material that is very weak in tension. Cement and concrete structures are designed so that the material is used predominantly in compression, with the aid of steel reinforcement or pre-stressing where appropriate. The great bulk of the literature on the mechanical properties of cement is concerned, therefore, with behaviour under compressive stresses (for a recent review see Sereda *et al.* 1980). The information summarized in the current paper is therefore of peripheral interest to the traditional uses of cement and concrete.

The recent development of macro-defect-free (MDF) cement with greatly enhanced mechanical properties (for example bend strengths of 200 MPa) shows that cement can now be considered for a range of applications, hitherto impracticable, where the material can be used under significant tensile stresses. A straightforward application of fracture mechanics and the Griffith equation (1) indicates that the MDF cements have increased values (about twice) for the Young modulus and surface energy and, of major significance, a greatly reduced (by about an order of magnitude) flaw size (Birchall *et al.* 1981, 1982). The detailed explanation must be more complex because the fracture toughness of cement is not a material constant and is significantly affected by specimen geometry and test conditions (Bailey & Higgins 1981; Alford *et al.* 1982).

In conclusion, based on the state of the art for ceramics, a number of comments are listed below that might be considered when applying fracture mechanics to the tensile behaviour of cements. It is hoped that these will stimulate improved understanding for cements.

(1) The application of fracture mechanics to cements is more complex than for ceramics.

(2) A statistical scatter in the strength of cements is expected; Weibull statistics can probably be applied.

(3) Subcritical crack growth under stress is expected for cements.

(4) The properties of cements (in the absence of stress) will change with time in sympathy with microstructural developments.

(5) Significant variation in the fracture toughness of cement with crack length should be expected.

(6) Some measure of crack healing could occur through microstructural changes.

(7) The concept of a flaw size in a very porous material like cement is complex and will make interpretation of strength difficult.

(8) The deliberate introduction of a fine dispersion of cracks, for example by the growth of fine particles during ageing, could lead to useful enhancements in properties through highly localized microcracking.

(9) Design data for engineering application of new cements require development. These could be based on the same principles as those for ceramics, but behaviour is likely to be more complicated.

(10) A considerable amount of research effort is required on the fracture mechanics of cement to improve understanding and stimulate engineering uses for the material.

References

Alford, N. McN., Groves, G. W. & Double, D. D. 1982 *Cem. Concr. Res.* **12**, 349.

Bailey, J. E. & Higgins, D. D. 1981 *Nature, Lond.* **292**, 89.

Birchall, J. D., Howard, A. J. & Kendall, K. 1981 *Nature, Lond.* **289**, 388.

Birchall, J. D., Howard, A. J. & Kendall, K. 1982 *Proc. Br. ceram. Soc.* **32**, 25.

Davidge, R. W. 1979 *Mechanical behaviour of ceramics*. Cambridge University Press.

Davidge, R. W., McLaren, J. R. & Tappin, G. 1973 *J. Mater. Sci.* **8**, 1699.

Griffith, A. A. 1920 *Phil. Trans. R. Soc. Lond.* A **221**, 163.

Gumbel, E. J. 1958 *Statistics of extremes*. New York: Columbia University Press.

McLaren, J. R., Davidge, R. W., Cotton, D. C. L., Haywood, S. A. & Robson, M. E. 1978 *Proc. Br. ceram. Soc.* **26**, 67.

Okada, T. & Sines, G. 1983 *J. Am. ceram. Soc.* **66**. (In the press).

Pankow, D. H. & Finnie, I. 1979 In *Mechanical behaviour of materials* (ed. K. J. Miller & R. F. Smith), vol. 3, p. 67. New York: Pergamon Press.

Sereda, P. J., Feldman, R. F. & Ramichandran, V. S. 1980 In *Proc. 7th Int. Congr. Chem. Cem., Paris*, vol. 1, sub-theme 6–1 p. 4. Paris: Éditions Septima.

Weibull, W. 1951 *J. appl. Mech.* **18**, 293.

Wiederhorn, S. M. 1974 In *Ceramics for high performance applications* (ed. J. J. Burke *et al.*), p. 633. Chestnut Hill, Mass.: Brook Hill.

Wiederhorn, S. M., Fuller, E. R. & Thomson, R. 1980 *Metal Sci.* **14**, 450.

Discussion

H. F. W. TAYLOR (*University of Aberdeen, U.K.*). In the first mechanism of zirconia toughening that Dr Davidge mentioned, how rapid must the transformation be for it to be effective? One can think of several possible expansive processes in a cement paste, but I wonder if any of them would occur quickly enough.

R. W. DAVIDGE. In the first toughening mechanism the transformation takes place through a martensitic reaction; this very rapid response is essential, otherwise at high strain rates the crack could propagate before the benefits of toughening have been obtained. The possible expansive processes in cement would not be of the above type, but there could be benefits through the second type of toughening mechanism involving a fine dispersion of microcracks.

J. E. BAILEY (*University of Surrey, U.K.*). Dr Davidge drew attention to subcritical crack growth problems in fracture mechanics. It is perhaps relevant to point out that rather tortuous subcritical crack growth has been observed in notched three-point bend tests on cement paste, indicating that stress concentrations are reduced and that cracks, however sharp, are effectively blunted by the cement microstructure.

R. W. DAVIDGE. Professor Bailey's observations indicate that the fracture toughness of cement should increase quite markedly with increase in crack length. Toughness values measured from fracture mechanics type specimens with large preformed notches are thus probably not applicable to the initial propagation of the inherent flaws.

M. S. T. PRICE (*UKAEA, AEE, Winfrith, Dorset, U.K.*). One point that Dr Davidge has not had time to mention is that there is a problem with an inherently granular material of relating the results of strength measurement on test specimens to the bulk behaviour. The problem is rendered more difficult if non-uniform stress conditions are used.

A material that I would suggest has important analogies with cement is polycrystalline graphite. It is granular and extensively flawed. The main reason for pointing to the behaviour of graphite is the considerable study of the material for structural uses in nuclear and aerospace application. See for example a review such as 'Fracture in graphite' by M. S. T. Price, presented at The Sixth London International Carbon and Graphite Conference, Society of Chemical Industry, London, 20–24 September 1982.

The way forward must take due account of microstructure and take advantage of the power of finite element computational techniques.

R. W. DAVIDGE. For ceramics, it is relatively straightforward to use the principles of fracture mechanics and statistical variations in strength to relate the strengths of specimens with different sizes or stress distributions. Whether such principles can be applied to cement materials will depend on much more detailed information about their basic fracture behaviour.

J. SKALNY (*Martin Marietta Laboratories, Baltimore, Maryland, U.S.A.*). Are there any other materials that could be used to replace zirconia (ZrO_2) in toughening ceramics? What are the chances of using this toughening mechanism for a porous material like hydrated cement?

R. W. DAVIDGE. The only other material found to toughen ceramics, as observed with zirconia, is the chemically related but much more expensive HfO_2. (My replies to Dr Alford's comments are also relevant.)

N. McN. ALFORD (*I.C.I., Runcorn, Cheshire, U.K.*). Dr Davidge suggested that the incorporation of reactive particles might be a means of toughening cement. In zirconia-toughened ceramics it is still not clear whether microcrack formation causes toughening as Dr Davidge suggested, or whether the martensitic transformation itself dissipates strain energy. I would suggest that the benefits of incorporation of reactive particles would merely cause a reduction in strength and modulus, and benefits to toughness would be minimal if any.

R. W. DAVIDGE. I agree that incorporation of reactive particles into cement could cause deterioration of several mechanical properties. However, I believe that a fine dispersion of particles to produce localized cracking of dimensions rather less than the inherent flaws could in principle give useful improvements in toughness. It would clearly be useful to check these ideas experimentally.

C. D. POMEROY (*Cement and Concrete Association, Wexham Springs, Slough, U.K.*). Cement hydrates comprise many different components, which will have different toughnesses. How does this affect the application of fracture mechanics and the production of crack growth? Would crack arrest by tougher components significantly influence the fracture mechanism?

R. W. DAVIDGE. It is not so much the individual toughness of the component phases of concrete that affect fracture behaviour, but the various complex ways in which the components react with each other. Of particular importance are differences in thermal expansion behaviour and n elastic properties.

Phil. Trans. R. Soc. Lond. A **310**, 127–138 (1983) [127]
Printed in Great Britain

Improvements in the early strength properties of Portland cement

By G. K. Moir
Research Division, Blue Circle Technical, Blue Circle Industries p.l.c., London Road, Greenhithe, Kent, DA9 9JQ, U.K.

The fuel savings that can be achieved in the manufacture of Portland cement clinker by means of mineralizers are relatively small. Further, the most effective mineralizers such as fluorine are known to have adverse effect on cement quality. However, research has shown that the retention of controlled levels of mineralizers in Portland cement clinker can result in significant quality improvements and this may provide a greater incentive for their use. The mode of action of one particular mineralizer combination is discussed and results are illustrated with reference to clinkers prepared both in the laboratory and on the production scale.

1. Introduction

In cement making, the term mineralizer is used to describe substances that facilitate the sintering reactions that take place in the rotary kiln. These substances may be added to the raw mix or may be inherent in the raw materials at certain locations.

Two examples of mineralizers whose actions are well reported in the literature are fluorides, such as CaF_2, AlF_3 (Lea 1970; Eitel 1966) and calcium sulphate (Budnikov *et al.* 1966; Butt *et al.* 1974). The beneficial effect of fluorine upon combination is thought to arise both from a reduction in the temperature of initial melting (Heinmann *et al.* 1970) and from the formation of intermediate phases whose decomposition products readily form the clinker minerals (Gilioli *et al.* 1979; Gutt *et al.* 1970). Diffusion experiments (Johansen *et al.* 1979) have also indicated that the rate of C_3S formation may be accelerated by the increased width of the C_3S primary phase region in the C–S–CaF_2 system compared to that in the C–S–A–F system.

Similarly, the mineralizing action of SO_3 may be attributed to an increase in the quantity of the liquid phase essential for the sintering and homogenization of the clinker and to the formation of an intermediate sulphosilicate phase (Gutt *et al.* 1967).

Recent publications (Sarkar *et al.* 1980; Klemm *et al.* 1980; Kumar *et al.* 1981) have indicated an increased level of interest in the use of mineralizers, which can be attributed to a desire to reduce clinkering temperatures and achieve fuel savings. However, although the clinkering temperature required for satisfactory combination may be reduced by as much as 200 °C, when heat recuperation from the clinker cooler is considered, the potential energy savings are relatively small (Gardeik 1981); under 5 % for a 200 °C reduction in burning-zone temperature. Incentives to use mineralizers may therefore come from other considerations such as extended refractory life, reduced power consumption for raw mix preparation or improvements in cement quality.

From the literature, prospects of achieving the latter benefit with either fluorine or SO_3 appear slight. When incorporated in Portland cement clinker, fluorine is known to have an adverse effect on cement reactivity leading to extended setting times and reduced early strengths. It has also been reported that when SO_3 entered into solid solution with C_3S its hydraulic activity was impaired (Welch *et al.* 1960). When both SO_3 and fluorine were present the combined effect resulted in very low strengths.

However, during the development of a new type of expansive cement (Pollitt & Brown 1975) we discovered that provided fluorine (normally as CaF_2) and SO_3 (normally as $CaSO_4$) were used together and in the correct proportions (Murray & Brown 1978), cements could be produced with significantly improved strength properties.

This paper describes the development of high early strength mineralized cements. The potential for strength improvements, when controlled levels of mineralizers are retained within the clinker, is illustrated with reference to clinkers prepared both in the laboratory and on the production scale.

2. Discovery of potential for improved strength properties

The shrinkage of hydrated Portland cement paste on drying is an inherent characteristic, which combined with its low tensile strength, results in crack formation in many structures. To overcome this disadvantage, several types of expansive or shrinkage compensating cements have been developed, some of them deriving their expansive properties from the hydration of uncombined magnesia or lime. Generally, these suffer from the disadvantage that the lime or magnesia either hydrates prematurely before adequate hydrate structure has been developed, resulting in loss of expansive forces, or too late, resulting in cracking of the hardened concrete. To overcome these disadvantages a cement was developed that utilized the sulphate stabilization of C_2S first reported in the literature by workers at the Building Research Establishment (Welsh *et al.* 1968) but which had been recognized within the industry as a feature of sulphate rich material in clinker derived from kiln deposits. To produce this novel cement, sufficient calcium sulphate was added to a high silica ratio, low iron mix to stabilize C_2S to an intermediate temperature of *ca.* 1480 °C. Above this temperature, loss of stabilization coincident with desulphation resulted in the formation of C_3S with a characteristic microstructure in which free lime occurred as inclusions within the alite. This structure has the advantage that not only is the size of the free lime crystals closely controlled but also they are progressively exposed as the C_3S hydrates, and consequently the expansive forces, are generated in step with strength growth. Unlike alternative expansive cement systems in which the expansive reaction takes place after silicate hydration, and which are consequently fundamentally unsound, the encapsulated free lime approach yields a stable matrix.

During the development work, calcium fluoride was added to the sulphated raw mixes and it was quickly discovered that fluorine overcame the stabilizing effect of SO_3, resulting in a more normal type of clinker microstructure. However, the benefits were that clinkers with very high C_3S contents (*ca.* 80 %) could be combined at temperatures considerably lower than those required for typical ordinary Portland cement raw mixes and further, the cements gave exceptionally high strengths at 7 days and beyond. By controlling closely the levels of SO_3, fluorine and alkali metal oxides retained in the clinker, cements could be prepared with both early and late strengths considerably higher than those of current cements at the same specific surface area.

3. Combined mineralizing action of sulphate and fluoride

At the Research Division of Blue Circle Industries the ease with which a raw mix combines to give a satisfactory level of residual free lime (hereinafter referred to as combinability) is determined by a standard test. Small uniformly sized pellets of raw mix ground to a controlled fineness, or range of finenesses, are fired under a dynamic heating régime to a range of final temperatures. The final temperatures required to give free lime levels of 1% and 2% are determined by interpolation. These are referred to as the 1% and 2% free lime combinability temperatures.

Table 1. Influence of mineralizer proportions on combinability of works' raw mix

mineralizer addition to raw mix		clinker free lime (%)				combinability temperature for 2% free lime/°C
gypsum (%)	CaF_2 (%)	1350 °C	1400 °C	1450 °C	1500 °C	
nil	nil	—	8.3	5.5	3.4	1550
2.0	0.25	6.3	4.0	2.4	—	1470
2.0	0.50	5.0	3.7	2.0	—	1450
4.0	0.25	5.4	4.1	2.1	—	1460
4.0	0.50	5.2	3.1	1.7	—	1440
6.0	nil	5.2	4.0	2.3	—	1460
6.0	0.25	3.9	3.0	1.8	—	1440
6.0	0.50	2.4	1.7	1.6	—	1370

Table 1 illustrates the influence of the mineralizers calcium fluoride and gypsum upon the combinability of a works' raw mix. The lime saturation factor (l.s.f.), silica ratio (s.r.) and alumina ratio (a.r.) of this sample were 99%, 2.57 and 2.53 respectively, where

$$\text{l.s.f.} = \frac{CaO}{2.8\,SiO_2 + 1.2\,Al_2O_3 + 0.65\,Fe_2O_3},$$

$$\text{s.r.} = \frac{SiO_2}{Al_2O_3 + Fe_2O_3},$$

$$\text{a.r.} = \frac{Al_2O_3}{Fe_2O_3}.$$

Without the addition of mineralizers, combinability was difficult and required a final burning temperature of 1550 °C to reduce the level of uncombined lime to 2%. The addition of the mineralizer pair of gypsum and calcium fluoride enabled free lime levels of 2% to be attained at temperatures as low as 1370 °C. The lowest combinability temperature for 2% free lime was obtained with the highest level of mineralizer addition shown in table 1. Higher levels, particularly of calcium fluoride could lower this temperature still further but clinker fluorine levels of greater than *ca.* 0.25% are undesirable on quality grounds, as discussed in §5.

With the raw mix chosen to illustrate the effectiveness of the mineralizer pair, gypsum when used alone had a mineralizing action. However, as reported by Gutt & Smith (1968), and as exploited in the expansive cement described earlier, in certain compositions sulphate stabilization of C_2S can occur, leading to very difficult combinability. This effect is likely to occur with raw mixes low in iron oxide or alkali metal oxides, and particularly where there is a low content of K_2O relative to Na_2O.

In practice, gypsum additions to the raw mix are limited both by the SO_3 level of the final cement, which is restricted by national standards, and production aspects such as the formation of build-ups in the kiln or preheater.

On the basis of these and other results, an optimum mineralizer addition of 6 % gypsum and 0.5 % calcium fluoride was selected. This is appropriate to the method of clinker production employed in the laboratory. In large scale production kilns these levels would have to be modified on account of different levels of volatile retention according to process type.

Table 2. Results illustrating potential for increased C_3S contents

chemical parameters		mineralizer addition to raw mix		combinability temperature for 2 % free lime/°C	clinker free lime (1400 °C) (%)	clinker‡ C_3S (1400 °C) (%)
l.s.f. (%)	s.r.	gypsum (%)	CaF_2 (%)			
101	2.5	6.5	0.5	1360	1.3	58
102	2.9	6.5	0.5	1360	1.3	60
102	4.0	6.5	0.5	1380	1.6	65
102	5.8	6.5	0.5	1470	2.9	65
102	5.8†	6.5	0.5	1420	2.1	69
100	2.5	nil	nil	1550	8.3	38

† Raw feed residue reduced from 10 % to 5 % retained on 90 μm mesh.
‡ Calculated according to Bogue, taking both clinker free lime and SO_3 into account.

Table 2 indicates the manner in which a marked increase in clinker C_3S level can be achieved by using the mineralizer pair to achieve combination in mixes of high l.s.f. At a s.r. of 6 (the highest investigated) it was found advantageous to reduce the size of the raw mix residue from 10 % to 5 % retained on 90 μm. Even at this fineness, a non-mineralized mix with a s.r. of 6 and l.s.f. of 102 % required two firings at 1550 °C with intermediate crushing and grinding to bring the free lime level below 2 %. The results illustrate how the mineralizer pair enable high C_3S clinkers to be combined at temperatures lower than those required for most ordinary Portland cement (o.p.c.) clinkers.

4. Influence of mineralizers on reaction sequence during firing

To gain a better understanding of the reaction sequence and intermediate phases formed during the firing of mineralized compositions, a raw mix was fired to a range of intermediate temperatures at a heating rate of 10 °C/min and the phases present identified by X-ray diffraction. The chemical analysis of the raw mix studied is given in table 3. Potassium oxide was added as K_2SO_4 to control the K_2O/SO_3 ratio; the relevance of this to cement properties is discussed later.

The phases detected by powder X-ray diffraction and the temperature ranges over which they occurred are illustrated in figure 1. Note that the phases found coexisting are unlikely to represent equilibrium phase assemblages as the samples were not held at the final burning temperature. It should also be remembered that some of the phases will not have been present at the temperatures indicated, but formed during subsequent cooling.

C_3S was first detected after firing to 1250 °C. The appearance of this phase coincided, as expected, with a decrease in the quantity of C_2S but also in the disappearance of the fluoro-aluminate phase $C_{11}A_7 \cdot CaF_2$ and the appearance of C_3A. The disappearance of $C_{11}A_7 \cdot CaF_2$ may be due to removal of fluorine from the interstitial phase by C_3S, as suggested by Klemm *et al.* (1980).

Klein's compound $C_4A_3\bar{S}$ had a relatively wide temperature range of occurrence as an intermediate phase but was not detected in samples fired above 1350 °C. This is consistent with its absence from mineralized clinkers prepared both in the laboratory and on the production scale.

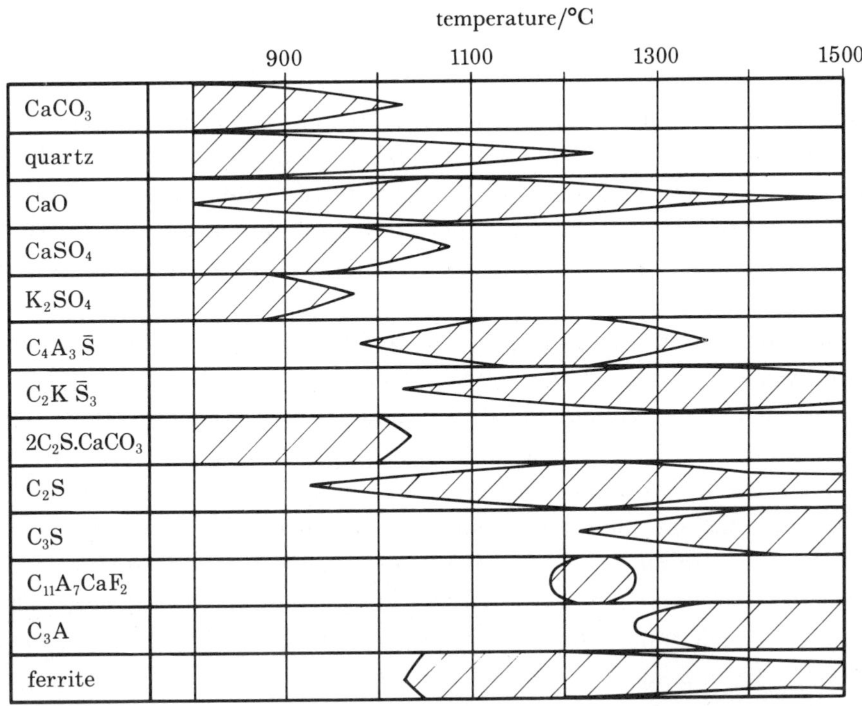

FIGURE 1. Temperature range of occurrence of phases in mineralized high C_3S composition. Samples were heated from 800 °C at a rate of 10 °C/min. Key: $C_4A_3\bar{S}$, Klein's compound; $C_2K\bar{S}_3$, calcium langbeinite; $2C_2S \cdot CaCO_3$, spurrite.

TABLE 3. RAW MIX COMPOSITION

	%		%
SiO_2	14.6	K_2O	0.78
Al_2O_3	1.8	Na_2O	0.02
Fe_2O_3	0.62	F_2	0.16
Mn_2O_3	0.04		
P_2O_5	0.04	l.s.f. (ignoring SO_3)	102
TiO_2	less than 0.01	s.r.	6.03
CaO	44.3	a.r.	2.90
MgO	0.27		
LOI	34.8	clinker potential C_3S (ignoring SO_3)	86.5
SO_3	2.3		

Raw mix residue 1 % with 90 μm mesh.

At temperatures above 1350 °C the phases present were C_3S, C_2S, C_3A, ferrite (C_4AF) and the double salt $C_2K\bar{S}_3$ known as calcium langbeinite. This salt plays an important role in controlling the early hydration reactions as it provides a source of soluble calcium sulphate.

The C_3S phase was invariably present as the rhombohedral form in these mineralized clinkers. This has been confirmed both by X-ray diffraction and by differential thermal analysis, which shows no inversion over the temperature range 20–1000 °C. In a study reported elsewhere (Aldous 1983) the levels of fluorine and Al_2O_3 necessary to stabilize this form have been investigated. An association between rhombohedral alite and enhanced strengths has already been detected from a study of production clinkers (Goudin *et al.* 1980). Our own investigations, using conduction calorimetry, indicate that the rhombohedral form as stabilized by Al_2O_3 and fluorine is significantly more reactive than either the monoclinic or triclinic forms.

5. Properties of laboratory prepared cements

To investigate fully the strength properties of these cements, sufficient quantities of clinkers were prepared in the laboratory for full scale testing according to BS 4550. The parameters investigated included l.s.f., s.r., a.r., and the level of SO_3, fluorine and alkali metal oxides retained in the clinker. The scope of this paper does not permit a discussion of all of the results obtained. Instead this will be restricted to the influence of the variables, a.r. and fluorine content.

The cements prepared in the investigation were all ground to a surface area of 450 $m^2\ kg^{-1}$ and used to make concrete specimens at a water/cement ratio (w/c ratio) of 0.55. This is lower than the normal 0.60 w/c ratio employed in standard tests, but was necessary because of the excellent workability characteristics of the mineralized cements, which arise in part from the integral set control provided by the calcium langbeinite present in the clinker.

The presence of sufficient calcium langbeinite in the clinkers for adequate set control has two advantages. First, no gypsum need be added at the grinding stage thus eliminating problems arising from variable degrees of gypsum dehydration. Second, the release of K^+ and SO_4^{2-} ions has a pronounced accelerating effect on early hydration. Normally, the accelerating action of alkali sulphate is accompanied by a marked reduction in strength at later ages. However, in these fluorinated clinkers the strength penalty accompanying an increase in clinker alkali content is relatively small and late strengths equal to or higher than those of current cements are easily attained. For comparison purposes, typical quality results for a U.K. rapid hardening cement (R.H.P.C.) ground to a surface area of 450 $m^2\ kg^{-1}$ are given in table 4.

Table 4. Typical strength levels given by U.K. rapid hardening cement (R.H.P.C.)

age/day	1	3	7	28
compressive strength/(N mm^{-2})	15	28	35	47

(Slump 35 mm.) Cement tested according to BS 4550 at a w/c ratio of 0.60.

One disadvantage associated with integral set control provided by calcium langbeinite is a tendency for coating to form on the mill lining and grinding media during cement grinding, leading to a reduction in grinding efficiency. This can be overcome by high efficiency closed circuit grinding.

Influence of a.r. ratio on strength properties

Results indicating the influence of a.r. at a s.r. of 4 are shown in table 5. The a.r. was found to have a marked influence on strength development at 8 h and 16 h but to have little effect beyond 1 day. Increases in early strength with increasing a.r. (i.e. increasing C_3A content) have

TABLE 5. INFLUENCE OF CLINKER a.r. RATIO ON QUALITY OF MINERALIZED CEMENTS

	clinker chemical parameters			
K_2O (%)	1.5	1.4	1.4	1.5
SO_3 (%)	3.9	3.4	3.3	4.0
F_2 (%)	0.17	0.18	0.17	1.17
a.r.	0.46	0.93	1.9	3.6
free lime (%)	1.6	1.4	1.8	2.6
	setting properties			
standard consistency water (%)	30.5	30.5	31.0	31.5
initial set/min	170	165	120	90
final set/min	215	220	145	115
	concrete strength†/(N mm^{-2})			
8 h	1.8	4.2	8.7	11.5
16 h	18.7	23.1	25.9	28.5
24 h	31.0	34.3	34.9	34.1
3 d	47.5	47.2	48.5	46.6
7 d	56.6	54.0	54.8	53.2
28 d	61.9	63.3	62.2	60.1
slump/mm	20	21	16	16

Raw mix residue < 2% with 90 μm mesh; l.s.f. 100%; s.r. 4; cement surface area 450 m^2 kg^{-1}.
† Water/cement ratio 0.55.

TABLE 6. INFLUENCE OF CLINKER FLUORINE CONTENT ON THE QUALITY OF MINERALIZED HIGH SILICA RATIO CEMENTS

		clinker chemical parameters					
K_2O (%)	1.0	1.2	1.4	1.5	1.4	1.2	1.5
SO_3 (%)	2.9	3.4	3.3	3.6	3.2	2.8	3.9
F_2 (%)	0.06	0.10	0.14	0.22	0.39	0.80	1.7
free lime (%)	3.3	2.0	1.7	1.7	1.6	1.5	1.1
		setting properties					
standard consistency water (%)	25.3	31.0	29.0	29.5	27.5	27.5	29.5
initial set/min	90	140	155	170	240	210	540
final set/min	110	175	205	220	310	260	660
			concrete strength†/(N mm^{-2})				
8 h	6.7	5.4	3.7	4.6	1.7	1.8	—
16 h	17.7	20.8	19.7	22.2	15.6	14.5	1.6
24 h	22.9	30.0	29.3	33.2	26.6	19.4	11.6
3 d	33.9	41.9	43.2	46.5	47.5	44.6	38.6
7 d	42.3	48.2	50.2	52.0	55.3	57.1	46.8
28 d	51.9	54.0	56.2	60.7	65.9	66.4	56.0

Raw mix residue < 2% with 90 μm mesh; l.s.f. 98%; s.r. 6; a.r. 2.5; cement surface area 450 m^2 kg^{-1}.
† Water/cement ratio 0.55.

been noted previously but not to the extent found in this study. At 8 h the strength given by the cement with an a.r. of 3.6 was similar to that given by current O.P.CS at 24 h.

Influence of fluorine content on strength properties

Results of strength tests are summarized in table 6. The reduction in hydraulic activity at early ages reported by previous workers was confirmed at high levels of fluoride addition. One day strengths, however, reached a maximum at a clinker fluorine level of *ca.* 0.25% and this improvement was greater than could be accounted for by the increased level of combination (as evidenced by clinker-free lime) arising from the mineralizing action of fluorine.

It is interesting to note that at later ages this optimum clinker flourine level become progressively higher. This may be analogous to the phenomenon of increased late strength commonly found when the rate of hydration at early ages is reduced by low temperature curing conditions or the use of retarders. Further evidence for the reduction in early hydraulic activity accompanying the increase in clinker fluorine levels comes from the progressive increase in setting times.

TABLE 7. QUALITY OF CEMENTS PREPARED FROM CLINKER PRODUCED IN A SUSPENSION PREHEATER KILN

trial number		1	2
clinker chemical parameters			
l.s.f. (%)		97.7	101.6
s.r.		3.7	3.4
a.f.		2.5	2.7
K_2O (%)		1.4	1.3
SO_3 (%)		4.0	3.3
F_2 (%)		0.19	0.25
free lime (%)		1.2	1.5
cement surface area/(m^2 kg^{-1})	488	322	450
setting properties			
standard consistency water (%)	29.5	26.0	—
initial set/min	160	160	—
final set/min	180	175	—
	concrete strength†/(N mm^{-2})		
8 h	3.2	—	—
16 h	19.1	—	—
24 h	29.4	21.7	25.7
3 d	40.6	33.7	39.0
7 d	46.2	39.3	45.1
28 d	53.3	46.4	51.0
slump/mm	55	50	55

† Water/cement ratio 0.60.

Even at the highest clinker fluorine level the phase $C_{11}A_7 \cdot CaF_2$ was not detected in the clinker. This phase was only found at burning temperatures below 1350 °C or when slow cooling was employed from higher temperatures.

6. PROPERTIES OF CEMENTS PREPARED FROM PRODUCTION CLINKERS

The potential strength improvements indicated by the laboratory work have been confirmed on the production scale with both wet process and suspension preheater plants. Concrete strength results obtained with clinkers sampled from two production trials on a suspension preheater plant are summarized in table 7. The strength results obtained with laboratory prepared clinkers were confirmed or even exceeded in some cases.

Figure 2 compares the strength development of cement prepared from mineralized clinker with that of current typical R.H.P.C.; both cements have a surface area of 450 m^2 kg^{-1}. Under standard laboratory test conditions at a fixed w/c ratio, compressive strengths up to twice those of current R.H.P.CS can be achieved. When concretes are gauged to constant slump the superior workability characteristics of the mineralized cements give even greater benefit to strength.

7. Summary and conclusions

The mineralizer pair of fluorine and SO_3 can be employed to produce cements with early strength properties markedly superior to those of current Portland cements. This is attributable partly to the higher alite content permitted by the mineralizing action but also to increased hydraulic activity associated with the occurrence of alite as the rhombodedral polymorph, and the presence of controlled levels of readily soluble alkali sulphate.

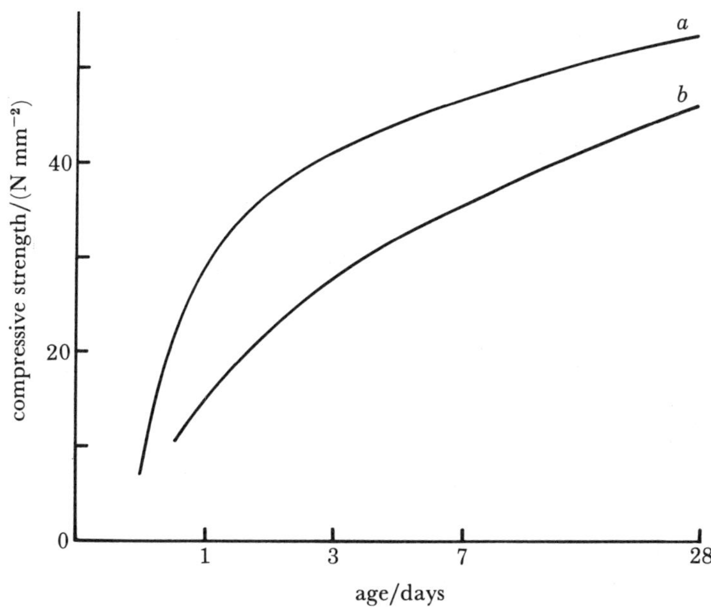

FIGURE 2. Strength development of mineralized cements prepared from production trial clinker. Tests made on 100 mm concrete cubes cured at 20 °C with mix 1:2.5:3.5:0.60. (*a*) Mineralized cement; (*b*) typical U.K. R.H.P.C.; both cements had surface area 450 m^2 kg^{-1}.

It is not suggested that cements of this type be employed as a replacement for current Portland cements in conventional concrete structures. Principal applications are expected to lie in two main areas. First, in the manufacture of cement based products where the high early strength development should enable considerable economies to be made, both in curing and in the turn of forming moulds. Second, in the manufacture of extended cements where the clinker is partially replaced by materials such as pulverised fuel ash or granulated blast-furnace slag. The high level of hydraulic activity of these mineralized clinkers overcomes one of the principal disadvantages of this type of cement, namely low early strength, and permits increased levels of clinker replacement and consequently increased energy savings.

I am grateful to Mr B. L. Morton of Research Division, Blue Circle Industries p.l.c., for many helpful discussions on this topic, and to Blue Circle Industries p.l.c. for permission to publish this paper.

References

Aldous, R. T. H. 1983 *Cem. Concr. Res.* **13**, 89.

Budnikov, P. P. & Kuznetsova, I. P. 1962 *S. appl. Chem.* **35** (5), 13–16.

Butt, Y. M. & Timashev, V. V. 1974 *6th International Congress on the Chemistry of Cement, Moscow.*

Eitel, W. 1966 In *Silicate science*, vol. 5, pp. 74–77. New York: Academic Press.

Gardiek, H. O. 1981 *Zem.-Kalk-Gips* **4**, 169–174.

Gilioli, C. & Mazzaza, F. 1979 *Cem. Concr. Res.* **9**, 295–302.

Gutt, W. & Osborne, G. J. 1970 *Trans. Br. ceram. Soc.* **69** (3), 125–129.

Gutt, W. & Smith, M. A. 1967 *Trans. Br. ceram. Soc.* **66** (11), 557–567.

Gutt, W. & Smith, M. A. 1968 *Trans. Br. ceram. Soc.* **67** (10), 487–510.

Heinmann, R. & Franke, W. 1970 *Ber. dt. keram, Ges.* **47** (3), 165–168.

Johansen, V. & Christensen, N. H. 1979 *Cem. Concr. Res.* **9**, 1–6.

Klemm, W. A. & Jawed, I. 1980 In *Proc. 7th Int. Congr. Chem. Cem., Paris*, vol. II, 150–155.

Kumar, S. S. & Kataria, S. S. 1981 *World Cem. Technol.* **40**, 279–285.

Lea, F. M. 1970 In *Chemistry of cement and concrete*, 3rd edn, pp. 156–157. London: E. Arnold.

Murray, R. J. & Brown, A. W. 1978 *U.K. Patent* no. 1498057.

Pollitt, H. W. W. & Brown, A. W. 1975 *U.K. Patent* no. 1406654.

Sarkar, A. K., Roy, D. M. & Smart, R. M. 1980 In *Proc. 7th Int. Congr. Chem. Cem., Paris*, vol. 2, 195–200. Paris: Éditions Septima.

Welch, J. H. & Gutt, W. 1968 In *4th Int. Symp. Chem. Cem., Washington*, vol. 1, pp. 59–68.

Discussion

H. F. W. Taylor (*University of Aberdeen, U.K.*). What is the physical state of the calcium langbeinite in the ground cement? Is it present as separate particles, or in the same particles as silicate and aluminate phases and perhaps surrounded by them? This would presumably affect its behaviour as a setting control agent.

G. K. Moir. In the mineralized clinker the calcium langbeinite occurs in the interstitial phase. Examination of cement with a scanning electron microscope indicates that the calcium langbeinite is widely distributed as a thin layer on the surface of the cement particles. It is certainly readily accessible to water and typically *ca.* 90 % of the total K_2O content of this type of cement is extracted in water by using the method described by Pollitt *et al.* (1968).

F. Massazza (*Italcementi S.p.A., Bergamo, Italy*). Sometimes we use mineralizers such as $CaSO_4$ and CaF_2 to improve burnability or other properties of clinker, but not to improve early strengths of cement. Do you believe that the cost of addition can be justified solely by the improvement in early strength?

G. K. Moir. The improvements in early strength reported will only be achieved if closely controlled levels of SO_3, fluorine and alkalis are retained in the clinker. If improvements in early strength have not been experienced then this may be because the levels retained in your clinkers are somewhat removed from the optimum.

Whether the cost of additives can be justified solely on cement quality considerations depends on market conditions and in particular whether a premium product can be sold at a premium price. But certainly in some circumstances the additional raw material cost will be justified.

J. E. Bailey (*University of Surrey, U.K.*). Are strength increases primarily due to increases in alite reactivity or do the aluminate phases play an increased role?

G. K. Moir. We consider that alite reactivity is a prime factor leading to increased strength, although for early strength, the alkali and sulphate content of the clinker is very important.

Cements with a high ratio of C_3A to ferrite phase show improved early strength as is indicated in table 5 of this paper. However, it is not known whether this arises from a direct contribution to strength made by aluminate hydrates or from the influence of the increased quality of aluminate phase upon alite hydration. The presence of fluorine in these clinkers also appears to promote solid solution of alumina in C_3A and this may have a role in improving strength properties.

S. Chanda (*University of Surrey, U.K.*). I would like to know if you have determined any other mechanical properties of these mineralized cements, such as modulus of rupture or the Young modulus, which may be of use from an engineering point of view?

G. K. Moir. These have not been determined to date. However, the products of hydration of the mineralized cements are essentially the same as those formed by ordinary Portland cement although produced more rapidly and in slightly different proportions, and we would not expect any significant difference in mechanical properties.

P. Pichat (*CDF Chimie, Paris, France*). First, what percentage of gypsum is added to the mineralized clinker? Second, can Dr Moir say at what industrial stage he is at, at the research or at the pilot stage? Third, has Dr Moir any experience with phosphogypsum, which contains about 0.6% fluorine?

G. K. Moir. First, no gypsum is usually added to the mineralized clinker. The best results are obtained with the integral set control provided by calcium langbeinite when no gypsum is added at the cement grinding stage.

Second, successful trials have been done on full scale production plants.

Third, when used as a retarder, phosphogypsum can give problems associated with prolonged setting times unless the level of soluble phosphate in the gypsum is maintained below certain levels. However, when added to the raw mix the phosphate is combined within the clinker and at the levels likely to be encountered has no discernible effect on cement properties. Any fluorine contributed by the gypsum when added to the raw mix is an additional bonus. Thus, mineralization according to the method outlined in our Patent (U.K. no. 1498057, 1978) permits the use of by-product gypsums, which are not suitable for use as conventional cement retarders.

Della M. Roy (*Materials Research Laboratory, The Pennsylvania State University, Pennsylvania, U.S.A.*). Concerning the hydration products from those cements having relatively high sulphate contents, are the long-term hydration products enriched in ettringite content relative to normal cements? This could affect their longer term performance.

G. K. Moir. As might be expected, whether ettringite persists depends on the ratio of SO_3 to the aluminate phases. In cements with a high s.r. (*ca.* 6) and SO_3 level of 3.5% we have found ettringite persisting at 90 days with no evidence of conversion to monosulphate. Concrete specimens have now been stored for up to 7 years with no evidence of unsoundness.

S. Sprung (*Research Institute of the German Cement Industry, Düsseldorf, F.R.G.*). (1) Does the addition of gypsum cause an increase in SO_2 emission?

(2) Does the addition of a fairly high amount of CaF_2 to the raw mix cause emission of HF?

(3) The addition of fluxes increase the proportion of melt in the clinker. Did Dr Moir observe or measure an increase of power consumption (kWh/t cement) during milling of the cement?

(4) Lowering the temperature in the sintering zone leads to a decrease in thermal efficiency of the clinker cooler. Did Dr Moir measure the heat consumption (kJ/kg) of the kiln system (kiln and preheater and cooler) with and without addition of fluxes?

G. K. Moir. (1) Not necessarily. SO_2 emission will depend upon process type and upon the lowering of burning zone temperatures and consequently of SO_3 volatilization, which the mineralizers permit.

(2) I would have to refer Dr Sprung here to his own paper (Sprung *et al.* 1968) in which he concludes that the exit gases of cement kilns cannot contain gaseous fluorides. We find very little volatilization of fluoride within the kiln system and any fluorine leaving the system does so as CaF_2 in which form it is insoluble and harmless.

(3) The mineralized clinkers described are generally softer and more porous than conventional clinkers. This is because of their generally lower conventional flux content as a result of the higher s.r's permitted, and also because of the lower burning temperatures.

However, when no gypsum is added during cement grinding a marked increase in specific power consumption can occur unless the mill is equipped with an efficient closed circuit milling system. This drop in grinding efficiency arises as a result of coating formed on the mill and media.

(4) During a continuous period of production of 8 days on a 35 t/h suspension preheater plant an overall net fuel saving of about 8% was achieved. Kiln temperature throughout was also slightly higher. No separate measure of cooler efficiency was made.

Reference

Pollitt, H. W. W. & Brown, A. W. 1968 In *5th International Symposium on the Chemistry of Cement, Tokyo*, pp. 322–333.

Phil. Trans. R. Soc. Lond. A **310**, 139–153 (1983) [139]
Printed in Great Britain

The relation between porosity, microstructure and strength, and the approach to advanced cement-based materials

By K. Kendall, A. J. Howard and J. D. Birchall, F.R.S.
I.C.I. New Science Group, P.O. Box 8, The Heath, Runcorn, Cheshire, U.K.

[Plate 1]

A theory is formulated to connect the strength of cement paste with its porosity. The theory shows that bending strength is largely dictated by the length of the largest pores, as in the Griffith (1920) model, but there is also an influence of the volume of porosity, which affects toughness through changing elastic modulus and fracture energy. Verification of this theory was achieved by observing the large pores in cement, and then relating bending strength to the measured defect length, modulus and fracture energy. The argument was proved by developing processes to remove the large pores from cement pastes, thereby raising the bending strength to 70 MPa. Further removal of colloidal pores gave a bending strength of 150 MPa and compression strength up to 300 MPa with improved toughness. Re-introduction of controlled pores into these macro-defect-free (MDF) cements allowed Feret's law (1897) to be explained.

1. Introduction

The purpose of this paper is to define, both theoretically and experimentally, the connection between the strength and porosity of hardened hydraulic cement pastes. Although this connection has been suggested many times previously, two outstanding problems have remained. First, there is the difficulty that ordinary cements are remarkably porous, an embarrassment that has traditionally been alleviated by polymer impregnation, by water reduction, or by high pressure compaction. Second, there is the lack of a theory relating cement strength to porosity; mathematical descriptions of cement strength have been empirically based.

Here we present a fundamental explanation of the relation between cement strength and porosity, starting from the Griffith (1920) model of brittle fracture, and postulating that cements contain two types of pore; colloidal pores, submicrometre in size, and 'crack-like' pores some millimetres in length. It is shown that the colloidal pores influence the elastic modulus and fracture energy of the cement, whereas the 'crack-like' pores initiate fracture and thus determine failure.

The experimental advance was in devising new processes for controlling the porosity of cement products. Addition of water-soluble polymer to the wet cement composition improved the plasticity of the slurry, allowing the material to be mixed in a novel way, which removed large pores. A further cold sintering step also permitted the colloidal porosity to be reduced. By studying these new cement products of controlled porosity, the theory connecting cement strength and porosity has been verified. The benefits of this approach were twofold. In the first place, Feret's (1897) original law of cement strength was explained and extended. Second, it has been demonstrated that strong hydraulic cements can be made by practical methods, thereby allowing them to compete with metals, plastics and ceramics in some applications.

2. The problem of weak cement products

Considering its ubiquity as an engineering material, conventional hardened cement is extremely weak. This weakness may be illustrated by contrasting the strength properties of hardened Portland cement paste with those of polymethylmethacrylate, sintered alumina, and steel (table 1). In fact, cement compares in strength to dry spaghetti. Sheets 100 mm square and 3 mm thick can be readily broken by bending between the fingers or by dropping from a few centimetres onto a hard surface.

Table 1. Strength properties of hardened Portland cement paste compared with polymethylmethacrylate (PMMA), sintered alumina, and steel

	cement	PMMA	alumina	steel
compressive strength/MPa	40	100	3000	500
flexural strength/MPa	7	130	500	700
fracture energy/(J m^{-2})	15	1000	30	10^5

That this weakness is associated with porosity may be proved by polymer impregnation of the cement. Organic monomer is soaked into a dry hardened cement and then polymerized (Swamy 1979). This leads to a doubling of modulus and quadrupling of tensile and compressive strengths, albeit from low values of 3 and 30 MPa respectively. The problem with this approach is the large volume, around 25 %, of organic material needed to fill the pores.

A more economical way of reducing cement porosity is to use less water in the mix. This can be achieved chemically by adding 1 % (by mass of cement) of surfactants such as sulphonated naphthalene or melamine formaldehyde condensates, which disperse the cement grains and allow the water content to be reduced from 0.39 to 0.26 w/c at equivalent rheology, to increase compressive strength from 41 to 74 MPa (Hattori 1978). Vibration has a similar effect on the dispersion of cement grains and allows comparable water reductions and strength improvements.

Alternatively, water can be squeezed out of the mix by applying pressure to the cement grains, pushing them closer together. For example, pressing cement paste in a hydraulic press at 10 MPa raises the bending strength from 10 to 20 MPa. Compaction at several hundred MPa can reduce porosity to 2 volume %, raising the bending strength to 60 MPa and compressive strength to 655 MPa (L'Hermite & Valenta 1937; Lawrence 1969; Roy & Gouda 1975). The difficulty with such brute methods is that only small articles of simple shape can be produced.

Having reviewed these experimental problems of porosity and strength, it is now necessary to study the theory.

3. The theoretical problem

The idea that porosity should determine strength is an old one; Lucretius wrote in the first century B.C. 'The more vacuum a thing contains within it, the more readily it yields...'. Early empirical studies of cement supported this view by correlating compressive weakness with the amount of water in the wet cement mix, as in Feret's law (1897). Feret connected compressive strength σ_c with the volume V_c of cement, V_w of water and V_a of air by the equation

$$\sigma_c = A[V_c/(V_c+V_w+V_a)]^2, \tag{1}$$

where A was a constant. This equation embodied the hydraulic paradox that water was essential to fluidize the cement and to provide hydrate reaction products, yet the more water was added, the weaker was the final hardened material. The strongest cements should be made with no water! (Abrams 1918).

Powers (1960) and Schiller (1958) explained this paradox by recognizing that the quantity of water in the mix controlled the volume porosity of the hard product. Since 1 cm^3 of Portland cement, on full hydration, converted to 2.2 cm^3 of hydrate, itself 28 % porous, it was clear that 1 cm^3 of water led to 0.513 cm^3 of porosity volume.

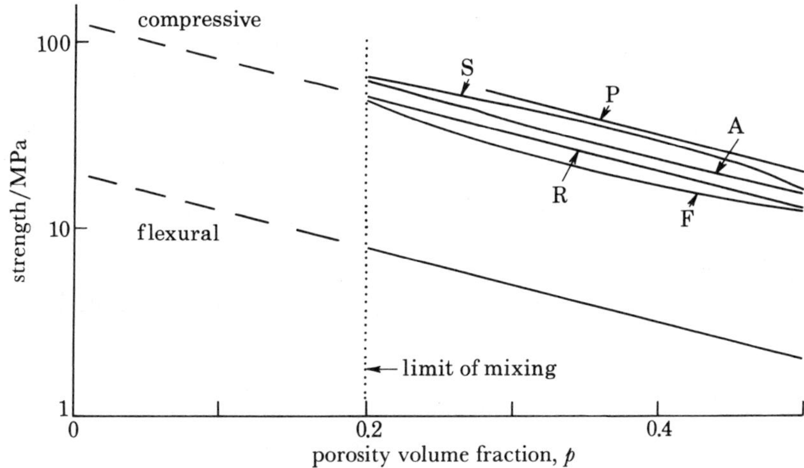

FIGURE 1. Empirical laws relating strength of cements to their volume porosity. (The compressive strength curves have been displaced vertically for clarity.) References: S, Schiller (1958); P, Powers (1960); A, Abrams (1918); R, Ryshkewitch (1953); F, Feret (1897).

Following this argument, it was the volume fraction of porosity p that caused strength to fall, and the several mathematical descriptions of cement strength could be plotted on a single graph (figure 1), which was a close approximation to Ryshkewitch's (1953) equation for the compressive strength of ceramics

$$\sigma_c = \sigma_0 \exp(-bp), \tag{2}$$

where σ_0 was the strength at zero porosity and b was a constant. Bending strength also followed this curve but was less than compressive strength by a factor of about seven.

In practice, porosities less than 0.2 are not attainable by merely mixing at such a low w/c ratio, but the curves may be extrapolated to zero porosity to estimate the intrinsic strength of cement. The conclusion is that ordinary cements are intrinsically weak with a maximum compressive strength of about 130 MPa and flexural strength of around 20 MPa.

The above logic, relating strength to *volume* porosity is well accepted in the literature, but does not fit experimental observations. Cements made by different processes can exhibit strengths that do not lie on the curves of figure 1. For example, cement that was frozen immediately after mixing, then thawed and hardened, gave a porosity volume increase of 3 %, but a strength decrease of 60 %. Birchall *et al.* (1981 *a*) showed that cement compositions could display the same volume of porosity but flexural strengths that varied by a factor of six.

The reason for this is that there is a particular kind of pore that has an enormous weakening effect. As Vitruvius observed 'Cracks make...bricks weak'. These crack-like pores need contribute very little to the total volume of pores. It is their length that is the important parameter, as Griffith (1920) demonstrated from the first law of thermodynamics. Tensile, or bending strength of brittle materials of Young modulus E and fracture energy R depend on the length c of a crack-like pore according to the equation

$$\sigma = (ER/\pi c)^{\frac{1}{2}}. \tag{3}$$

Application of equation (3) to cement products has been controversial. Kaplan (1961) concluded that it described the flexural fracture of notched concrete beams, but this was denied by Wright & Byrne (1964), who found that concrete was insensitive to notches. Other authors had difficulties with test methods (Kesler *et al.* 1972; Swartz *et al.* 1978) and some found that resistance to cracking varied with crack length and specimen size (Higgins & Bailey 1976; Brown 1972; Hillemeier & Hilsdorf 1977). Mindess (1981) has reviewed this issue.

For hardened cement pastes, Birchall *et al.* (1982*a*) showed that the Griffith theory did describe flexural strength of notched beams, but only for notches exceeding the natural flaw size, around 1 mm, in the material. Moreover, they pointed out that the key question was how to reconcile equation (2), which predicts a *volume* porosity dependence of strength and intrinsic weakness, with equation 3, which gives a pore *length* dependence and an apparently limitless rise in strength as pore length is reduced. This question was resolved by formulating a theory that explained both the pore volume and the pore size effects.

4. Theory of strength and porosity

The theory was based on the observation of two kinds of pore in hardened cement paste (figure 2, plate 1). There were colloidal pores trapped between the hydrate particles, at 2 nm in size too small to initiate cracking, but of sufficient volume to reduce the elastic modulus and fracture energy of the cement. Then there were crack-like pores some millimetres in length, long enough to initiate cracks at low stress levels, but contributing only about 10 % to the total volume fraction of porosity. Having recognized these two distinct types of pore, it was possible to imagine an ideal porous solid with two *independent* populations of pore; one population with volume p but no length; the other with length $2c$ but no volume (figure 3*a*).

A model of such a perfect material was porous silica glass (provided by Dr C. Howard, Pilkington Brothers) containing 48 % by volume of pores 10 nm in size, but no long cracks. Such a material was found to be strong, with a flexural strength of 60 MPa, proving that a large *volume* of pores does not necessarily cause low strength. However, its Young modulus E and fracture energy R were found to be reduced by the volume of pores, as shown by the results of figure 3*b*. When a crack-like pore of length 1 mm was cut into the porous glass with a saw, the bending strength fell dramatically to 5 MPa, reminiscent of cement strength, and the results were found to follow the Griffith (1920) curve (figure 3*c*). If the porosity volume of the porous silica was reduced by sintering, then the flexural strength again followed the Griffith relation when notched, but the curve was shifted to a higher level, because of the increased modulus and fracture energy.

To describe this behaviour mathematically, the Griffith energy balance approach was applied to the material (figure 3*a*) with the microporous glass treated as a continuum of the Young modulus (Hansen 1965; Helmuth & Turk 1966)

$$E = E_0(1-p)^3 \tag{4}$$

and fracture energy (Rice *et al.* 1978)

$$R = R_0 \exp(-kp), \tag{5}$$

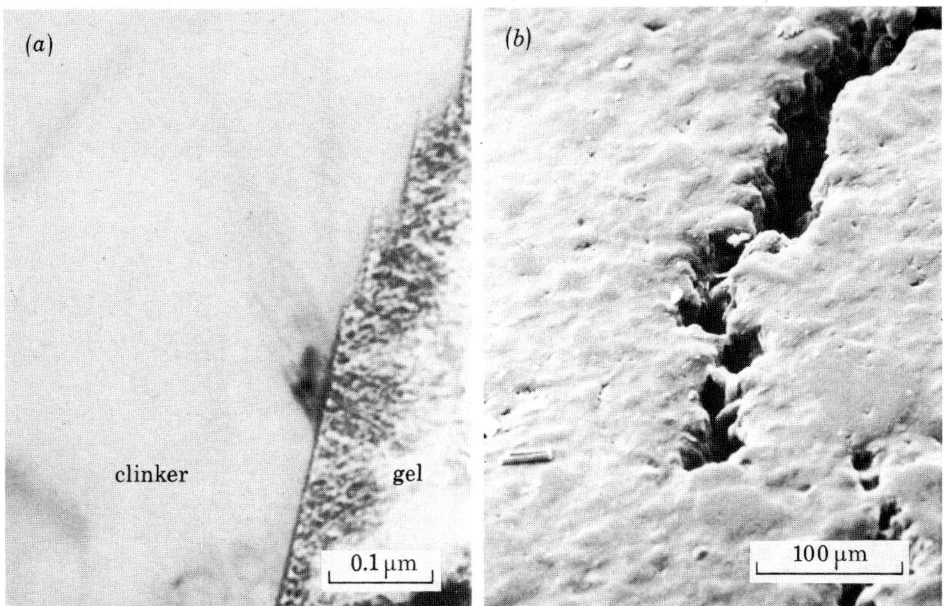

FIGURE 2. (*a*) Porous colloidal hydrate gel formed on an ion-thinned tricalcium silicate crystal (courtesy of Dr G. W. Groves). (*b*) Crack-like pore in polished cement paste; the cement grain residues can be seen protruding from the gel matrix.

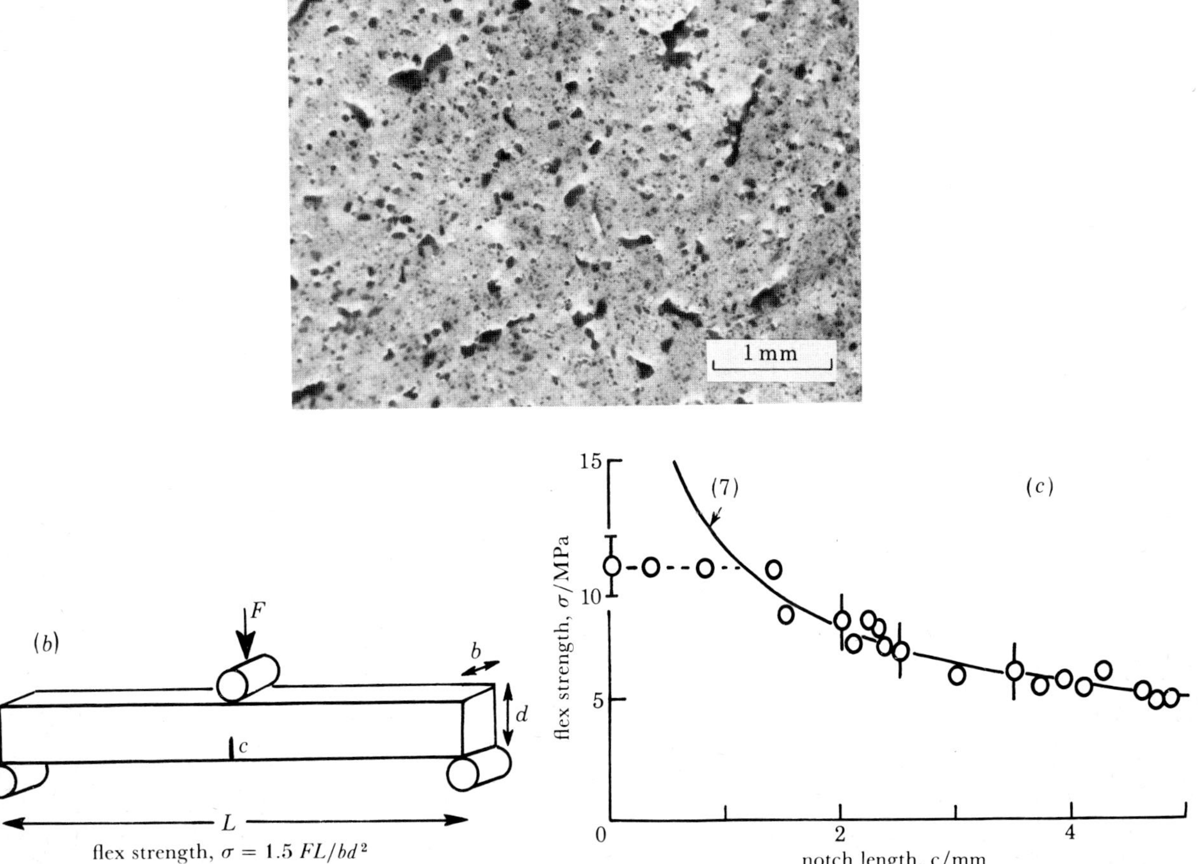

FIGURE 4. (*a*) Optical micrograph of a polished sample of Portland cement paste showing the large voids decorated black. (*b*) The flex testing arrangement. (*c*) Results for hardened Portland cement, w/c = 0.22. The curve derived from equation (7) is drawn as a solid line.

(*Facing p.* 142)

where E_0 and R_0 were the values at zero porosity volume. By calculating the energy U associated with a crack of length $2c$ (equivalent to an edge crack of length c) in an infinite sheet under a tensile stress σ, and applying the condition for energy conservation (given in Lawn & Wilshaw 1975, for example)

$$dU/dc = 0, \tag{6}$$

the criterion for crack extension was found to be

$$\sigma = \left[\frac{E_0 R_0 (1-p)^3 \exp(-kp)}{\pi c}\right]^{\frac{1}{2}}. \tag{7}$$

This equation was plotted in figure 3*c* to compare with the experimental results for porous silica glass, and was found to give reasonable agreement.

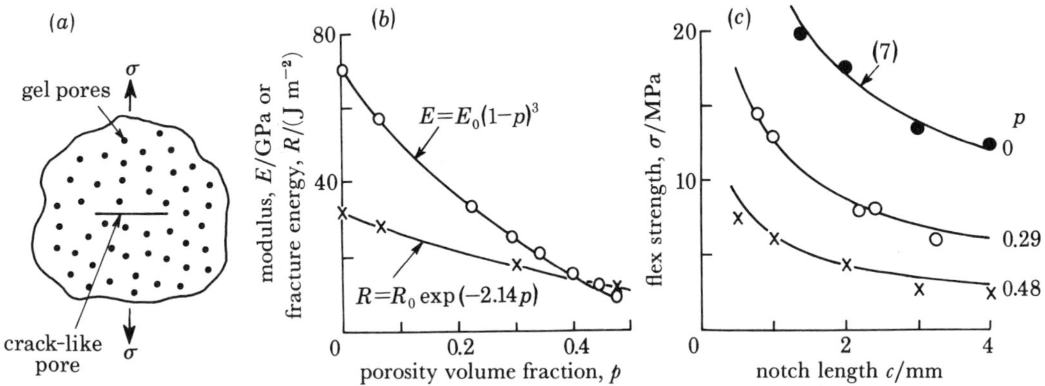

FIGURE 3. (*a*) Theoretical model showing two types of pore; the small gel pores of volume fraction p, and a long crack-like pore of length $2c$ causing fracture. (*b*) Variation of Young modulus E and fracture energy R with volume fraction of pores in porous silica glass. (*c*) Flexural strength as a function of notch length and pore volume for porous silica glass. This graph shows good agreement with equation (7), which is plotted as solid lines.

Equation (7), derived from thermodynamic considerations, is interesting for several reasons. It demonstrates that strength is a function of two independent variables, volume of porosity and length of a crack-like pore. Thus, the plotting of strength against pore volume p, a common error in the cement literature, is not sensible unless pore length c remains constant. Nor can strength be plotted against crack length unless porosity volume is constant. However, the main benefit of equation (7) is that it reveals two methods for strengthening porous materials such as hardened cement; the reduction of the volume of pores from 30 % to zero increases strength by a factor 2 (figure 3*c*); but the reduction in length of crack-like pores from 1 mm to 0.01 mm increases strength by a factor of 10. These predictions were tested experimentally on cement.

5. Experimental study of large crack-like pores

It is perhaps surprising that the long crack-like pores in hardened cement have not been much studied. One possible reason for this is that mercury intrusion porosimetry, the conventional method adopted by many workers (Winslow & Diamond 1970; Diamond 1973; Beaudoin 1979), does not give a correct description of large pores in cement. Several samples of cement paste were sectioned, polished, and decorated with carbon black to reveal the large pores under optical microscopy (figure 4*a*, plate 1). Measurements of several large pores gave a mean maximum pore length of 0.83 ± 0.17 mm and a volume fraction of around 10 % of pores exceeding 15 μm in length. When tested by mercury porosimetry, the volume of large pores was seriously underestimated, in one case by a factor 44 (table 2). The conclusion was that the large pores in cement were not detected by mercury, either because they were closed pores, or because the pores had small entrances and so were registered as pores of small size (Birchall *et al.* 1981 *a*). This conclusion has since been verified by others (Alford & Rahman 1981; Alford & Double 1982).

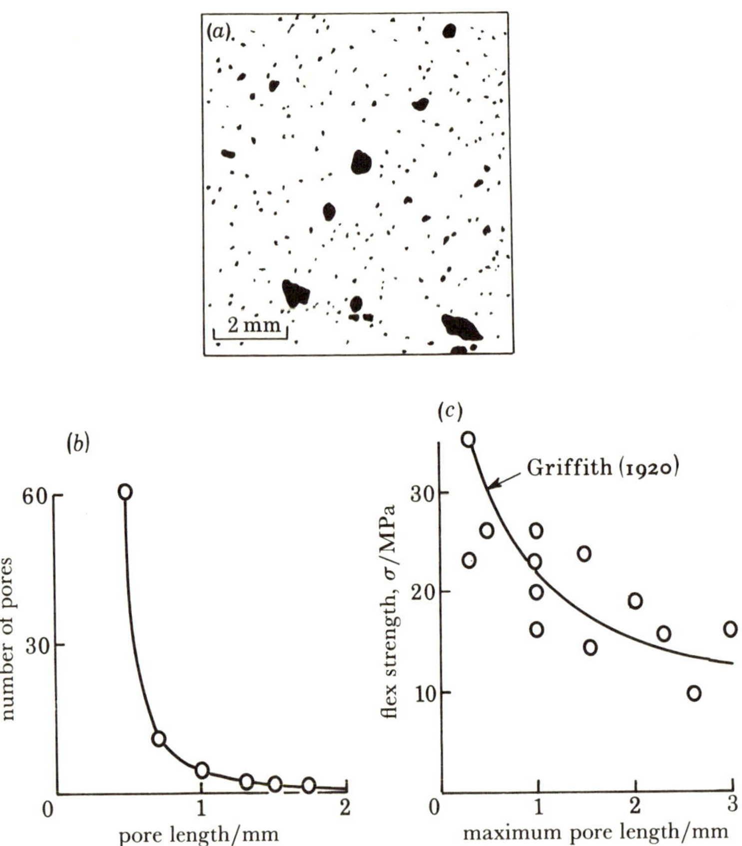

FIGURE 5. (*a*) Typical magiscan trace showing the distribution of flaws in a white cement surface (Secar 71, Lafarge, w/c = 0.35) cast against a smooth surface. (*b*) Magiscan plot of pore lengths over surface. (*c*) Bending strength dependence on length of natural flaws in white Portland cement (Snowcrete, Blue Circle, w/c = 0.25). The solid curve is drawn from Griffith's (1920) theory.

TABLE 2. VOLUME FRACTION OF LARGE PORES, GREATER THAN 15 μm IN LENGTH, MEASURED IN CEMENT SAMPLES BY TWO DIFFERENT METHODS; MERCURY POROSIMETRY, AND QUANTITATIVE MICROSCOPY

	volume fraction of pores longer than 15 μm	
sample number	mercury porosimetry	quantitative microscopy
1	0.0028	0.048
2	0.0011	0.048
3	0.0063	0.065
4	0.0097	0.114

Bending tests on notched cement samples were undertaken to see whether the long crack-like pores could be detected mechanically. A typical experimental arrangement is illustrated in figure 4 *b*, which shows a cast block of Portland cement paste, notched with a diamond saw, and loaded in flexure. For notches less than 1 mm in length, the flexural strength was not reduced (figure 5*c*), whereas notches longer than 1 mm caused the strength to diminish along the Griffith curve. Thus the natural crack-like pores in ordinary cement are about 1 mm long (Birchall *et al.* 1981 *a*).

In bending tests, the surface pores are dominant, but they cannot easily be seen on ordinary cement surfaces. To enhance pore visibility, white cement was cast against an optically smooth surface such as polyester film. After hardening, the film was removed and the cement surface was seen to be mottled when viewed in reflected light, the dark areas proved to be pores when probed with a needle. The size distribution of these pores was determined by image analysis and it was discovered that approximately 0.1 area fraction of the surface consisted of pores exceeding 0.25 mm in size, some up to 2 mm in length (figure 5). The strength of the samples was found to depend on the measured pore length in the stressed region, roughly fitting the Griffith (1920) theory. It was concluded that ordinary cast cement contains pores some millimetres in length, and that these pores limit the strength in bending. Alford (1981) has verified these conclusions.

The removal of these large natural pores from cements should result in a significant improvement in cement strength, even for materials of high volume porosity.

6. Origin and removal of long pores

Long pores in hardened cement originate from the hydraulic process. The conventional process of mixing cement powder with water and casting into a mould leaves large air bubbles and packing defects in the final product. Air bubbles cannot fill with hydrate gel because there is no water to transport the material. Packing defects contain water but insufficient cement grains to fill the space with gel, thus leaving a gap in the dried product.

Two observations suffice to show why the ordinary process leads to such voids. First, consider the packing of the dry cement grains. A bag of cement contains only 30 % volume fraction of cement; the rest is air. This observation shows that the cement grains are flocculated, stuck to each other to give a bonded open structure with large voids and a distinct yield point. Wetting the grains reduces the flocculation, but not to zero. Next, consider applying a mechanical force to the wet cement mix, as in the apparatus of figure 6*a*. Cement paste is often said to be plastic or 'Bingham' in its flow behaviour (Newman 1966; Helmuth 1980), but it differs from plasticine, which fits the mathematical theory of plasticity for this geometry (Hill 1950)

$$P = \tfrac{1}{4}Y(3+w/h), \tag{8}$$

where P is the pressure applied to the ram cylinder of diameter w acting on a thickness h of plastic material (figure 6*b*). Instead, cement paste gave a rapid increase in resistance P as the gap h was reduced, and this was explained by frictional interaction between cement grains sliding across each other, giving

$$P = \tfrac{1}{4}(Y+\mu P)\,(3+w/h), \tag{9}$$

where μ was the friction coefficient as in the behaviour of soils (Bolton 1979). Thus there are two forces acting on grains, leading to large pores; the attractive force between grains, described by the yield pressure Y, and the frictional force, described by μ, which prevents the grains from moving into their most close packed positions when mixed.

Addition of water to cement, or vibration, or use of superplasticizers, reduced the yield pressure but not the friction coefficient. However, addition of water-soluble polymers such as hydroxypropylmethylcellulose (HPMC), poly(acrylamide), or poly(vinylalcohol/acetate) reduced both yield pressure and friction (figure 6*c*). Such mixtures did behave as plastic bodies and could be worked by rolling, extrusion, and dough mixing (Downing *et al.* 1976; Birchall *et al.* 1981 *b–d*, 1982 *b*).

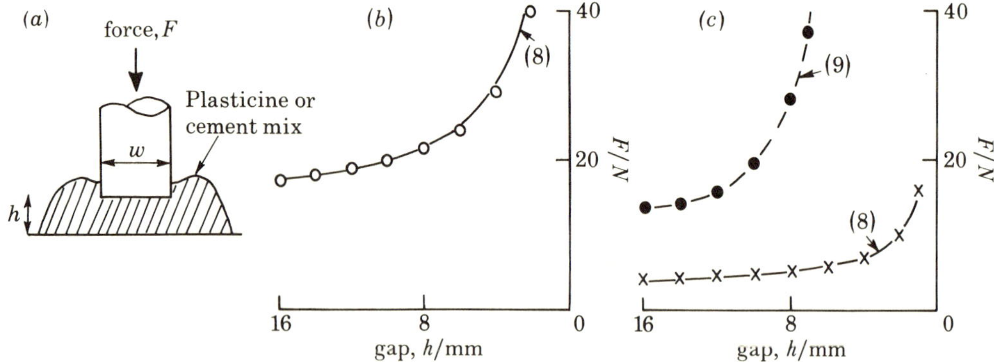

FIGURE 6. (*a*) Apparatus for determining the plasticity of pastes. A cylindrical ram was pressed into the paste at constant speed and the force measured as function of gap *h*. (*b*) Graph to show that results for Plasticine (o) fit the plastic theory (solid line drawn from equation 8). (*c*) Ordinary cement paste results (•) do not fit the plastic theory but fit equation (9) (dashed line), which assumes a friction coefficient between the grains (μ = 0.7). When 3 % of poly(vinylalcohol/acetate) was added the results (crosses) fitted equation 8 (solid line).

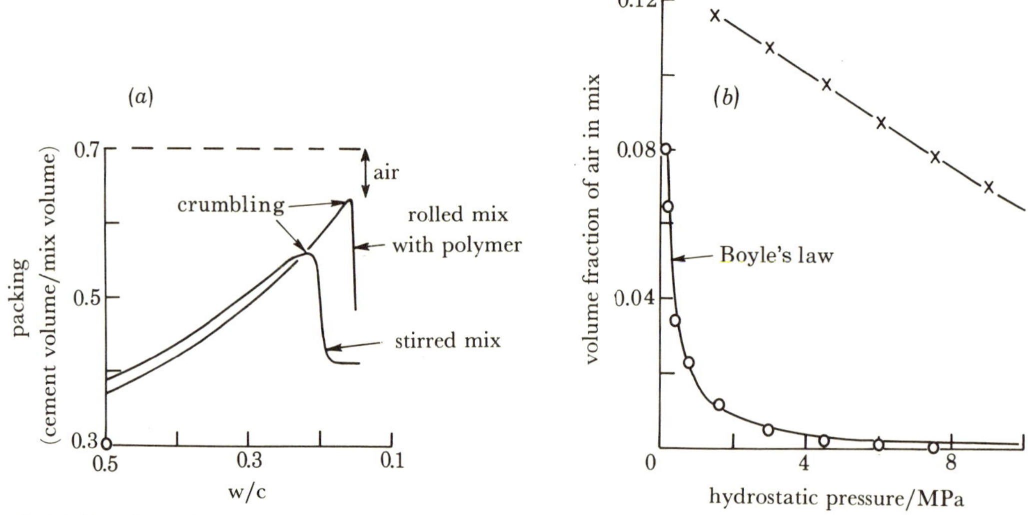

FIGURE 7. (*a*) Graph to show that increasing the packing of cement grains in a stirred mix by reducing water is limited by crumbling of the mix at about 0.22 w/c. But when polymer was added and the mix was rolled, the mix did not crumble until 0.15 liquid/cement ratio. The origin is the packing of dry cement powder. (*b*) Air in an ordinary cement/water mix (100/19) was not easily pressed out, but in a mix containing polymer (100/16/3HPMC), air was compressed according to Boyle's law. *Key*: crosses, cement/water mix results; circles, cement/water/polymer mix results.

The benefits of this dramatic change in rheology were shown in figure 7*a*. Whereas ordinary stirring of a mixture of water and cement produced a maximum packing of 0.56 before crumbling occurred, roll-mixing of water, cement and polymer gave 0.63 packing. Both such mixes contained nearly 10 % by volume of air. But whereas the air was only compressed with difficulty out of the ordinary cement, it was readily compressed according to Boyle's law from the polymer mix, owing to lower friction. The air almost disappeared at 5 MPa pressure, giving a packing of 0.7 (figure 7*b*).

It should be emphasized that polymer addition, by itself, did not result in good packing, nor high strength. In fact, merely adding polymer to a stirred mix at the same volume packing resulted in crumbling and low strength (table 3). The best results were obtained by polymer addition, dough mixing, and air removal. By these three operations, the microstructure of the hardened cement was much changed, with no pores greater than 15 μm long when it was polished, stained with carbon black, and viewed in the optical microscope. For this reason, these new materials were called macro-defect-free (MDF) cements.

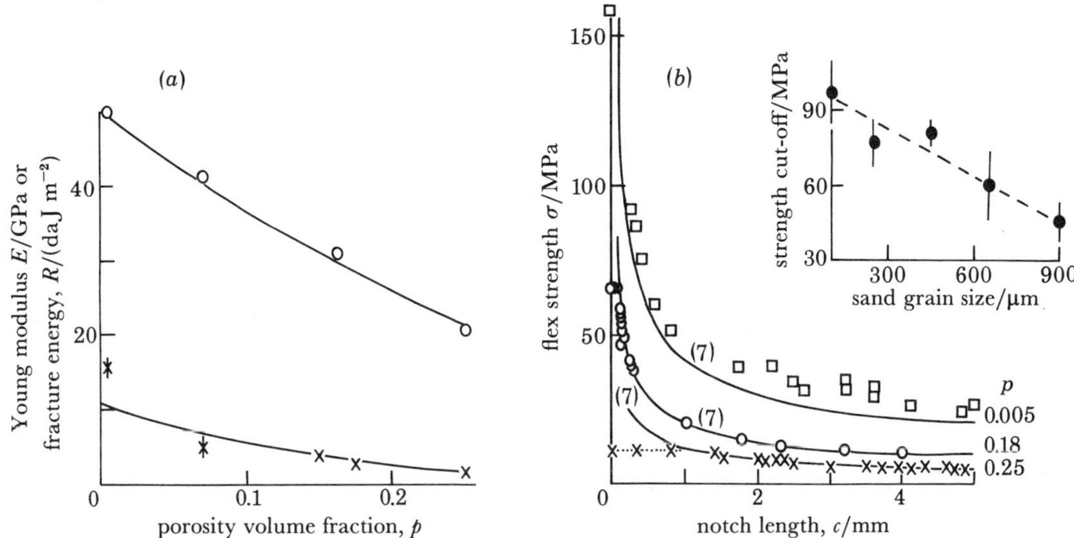

FIGURE 8. (*a*) Variation in Young modulus and fracture energy of cement with porosity *p*. (*b*) Flexural strength plotted against notch length for three different porosity cements compared with equation (7) (solid lines). Inset: change in the strength cut-off with grain size of sand particles sprinkled on the mix before pressing.

TABLE 3. RESULTS SHOWING THE THREE STEPS REQUIRED TO MAKE MDF CEMENT WITH CALCIUM ALUMINATE (SECAR 71, LAFARGE) AND POLYMER (CELACOL 15000 DS, CELANESE)

(Neither polymer nor pressure were in themselves sufficient to give high strength.)

mix	flex strength/MPa
control (100c/22w†, stirred, hand laid)	14 ± 2
control plus polymer (100c/18w/4p, stirred, hand laid)	1 (crumbly)
control plus pressure (2.5 MPa)	20 ± 5
control plus pressure (10 MPa)	25 ± 8
control, polymer and pressure (100c/18w/4p, 10 MPa)	27 ± 2
control, polymer plus dough mix (100c/18w/4p, twin-roll mill)	40 ± 2
control, polymer, dough mix, press set (100c/18w/4p, roll, pressure set at 5 MPa)	55 ± 3

† Note: c, cement; w, water; p, polymer.

When notched and tested in bending, MDF cement fitted equation (7) (figure 8*b*, middle line), but was now distinctly different from ordinary cast cement (figure 8*b*, bottom line). First, because less water was used and some air was removed, the volume porosity was reduced, so the strength results lay on a higher Griffith curve. Second, and more important, the results fitted the curve at smaller notch depths, down to 0.1 mm, because the large natural pores had been eliminated. Unnotched bending strength thus rose to 66 MPa. Although the visible pores were reduced far below the 0.1 mm level, the strength did not rise further. It was suggested that flaws other than pores were becoming important at this stage. Significantly, the largest cement grains were around 0.1 mm diameter, so that elastic mismatch or poor grain bonding may have become significant. Placing sand grains of various sizes in the surface of a dough before pressing, progressively dropped the strength of the material (figure 8, inset).

7. Origin and removal of colloidal pores

The strength increases above were obtained by the removal of the long crack-like pores, a significant volume of colloidal pores remaining in the cement product, 0.18 volume fraction. These colloidal pores stemmed from the gel hydration product, which has been thought tc be intrinsically porous to allow osmosis and continued hydration (Double *et al.* 1978; Birchall *et al.* 1978, 1980). Yet, in certain circumstances, colloidal pores can be removed from hydraulic systems. For example, silica gel on sintering at 1000 °C shrinks as the gel particles fuse together to eliminate porosity. Similarly, a latex of colloidal polymer particles shrinks at room temperature as drying occurs to pull the particles into a non-porous mass (Kendall 1980, 1982).

We have found a way of doing this room temperature sintering operation on hydraulic cements, by using particular combinations of water-soluble polymer and cement (Birchall *et al.* 1982*b*). For example, 100 g of calcium aluminate powder (Secar 71, Lafarge) was mixed with 7 g of poly(vinylalcohol/acetate) (KH 17s, Nippon Gohsei) and 11 g of water. When stirred, this mixture formed a crumbly composition, which was twin-roll milled to a stiff dough. This dough was found to exhibit a rheology quite different from doughs made with other cements and polymers. The shear stress doubled for a tenfold increase in shear rate, whereas Portland cement mixed with poly(acrylamide) and water gave roughly constant shear stress as the shear rate was increased.

The dough was press moulded at 5 MPa to remove air bubbles, then hydrated in the mould by heating at 80 °C for 10 minutes. After removal from the mould, the hydrated sheet was dried at 80 °C in an oven, and it was observed that significant shrinkage occurred, up to 3 % linear or 9 % volume. This was thought to correspond to the coalescence of the hydrate precipitate and the collapse of the colloidal porosity in the hardened cement, equivalent to sintering the material. Polished cross sections of this product showed no large pores. The material was thus proved to be a special case of macro-defect-free cement.

When tested for porosity by soaking in organic fluid, this material was shown to be much less porous than normal MDF cement. Usually, MDF cement absorbed between 10 % and 15 % by volume of fluid, whereas the shrunk material absorbed only 0.5 %, suggesting a low volume porosity. Moreover, the Young modulus was 50 GPa (figure 8*a*) consistent with a porosity below 0.01.

The material was notched and flex tested as before (figure 8*b*, top line). Again the results followed equation 7 with a cut-off around 0.1 mm notch length, giving a flex strength of around 150 MPa. The fracture energy was somewhat higher than expected from equation (7), 160 J m^{-2} compared with the theoretical prediction of 106 J m^{-2}, and this suggested that the polymer additive was making some contribution to crack resistance. However, it may be seen from figure 8 that equation 7 gives a reasonable description of the strength properties of the three cements tested, with porosity as the only variable.

8. Feret's (1897) law

It was now possible to re-interpret Feret's (1897) original empirical description of cement strength embodied in equation 1, while recognizing that almost half the water volume in Feret's equation is converted into pores. Thus the strength is

$$\sigma = A[V_c/(V_c+V_w+V_a)]^2 = A(1-1.95p)^2 \quad \text{when} \quad V_a = 0. \tag{10}$$

This expression was plotted in figure 9 to compare with equation (7) calculated for several notch sizes c. Agreement was good when the notch length was taken to be 2 mm, and $A = 30.2$. Thus it seems that Feret's law describes the strength of cements at constant flaw size, that is the toughness of cement. By some happy accident, when cements were made by the conventional mixing

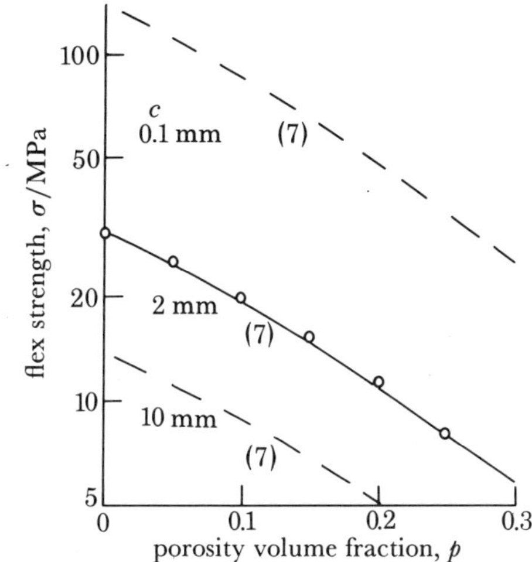

Figure 9. Comparison between Feret's (1897) law (○) and equation (7) (dashed and solid curves) plotted for several different flaw sizes c. Feret's law fits only at one flaw size; thus it describes the *toughness* of cement and not its *strength*.

process, the flaw size must have remained roughly constant when different water contents were used. Feret's law is therefore only partly true because it does not take into account the variation of flaw size c, which can be obtained by using different processes to mix cement. For example, freezing the mix would introduce long flaws and would give the lower curve in figure 9, whereas the MDF process would give the top curve. The true description of cement strength is equation (7), which plots onto figure 9 as a family of Feret curves, each corresponding to a constant flaw size.

9. Conclusions

A theory has been proposed to explain the connection between bending strength and porosity of hardened cement pastes. This theory indicates that the main influence on strength is the presence of macro-defects, which can initiate fracture in the material. Microscopic examination of ordinary cements has shown that such macro-defects may be identified as large pores, around 1 mm in length, stemming from imperfections in the mixing process. Improvements in mixing by (i) adding a polymer to make the mix plastic, (ii) employing high shear mixers and (iii) removing trapped air, produced a macro-defect-free product with a high bending strength of 70 MPa, even though the porosity volume remained at a high level. Thus, pore volume is not the major determinant of cement strength as suggested by previous workers. However, the volume of colloidal pores was found to influence strength by changing the elastic modulus and fracture energy of the material. By reducing the porosity volume to low levels after removing the macro-defects, the bending strength was further increased to 150 MPa, without the need for reinforcement or high-pressure compaction. The results were in fair agreement with the predictions of the theory.

Table 4. Improved properties of macro-defect-free (MDF) cement compared with ordinary cement

	ordinary cement	MDF cement
flex strength/MPa	10	40–150
Young modulus/GPa	20	35–50
compressive strength/MPa	40	100–300
fracture energy/($J\ m^{-2}$)	20	40–200
electrical breakdown/($kV\ mm^{-1}$)	0.1	6–13
gas permeability/($mol\ m^{-1}\ s^{-1}\ Pa^{-1}$).	10^{-12}	10^{-16}

The improved values of strength and toughness exhibited by macro-defect-free cements (table 4) should allow these materials to be employed in hitherto unexpected and advanced applications.

References

Abrams, D. 1918 *Bull. Univ. Chicago struct. Mater. Res. Lab. Lewis Inst. Chicago, Bull.* no. 1.
Alford, N. McN. 1981 *Cem. Concr. Res.* **11**, 605–610.
Alford, N. McN. & Double, D. D. 1982 In *Adsorption at the gas–solid and liquid–solid interface* (ed. J. Rouquerol & K. S. W. Sing), pp. 259–268. Amsterdam: Elsevier.
Alford, N. McN. & Rahman, A. A. 1981 *J. Mater. Sci.* **16**, 3105–3114.
Beaudoin, J. J. 1979 *Cem. Concr. Res.* **9**, 771–781.
Birchall, J. D., Howard, A. J. & Bailey, J. E. 1978 *Proc. R. Soc. Lond.* A **360**, 445–453.
Birchall, J. D., Howard, A. J. & Double, D. D. 1980 *Cem. Concr. Res.* **10**, 145–155.
Birchall, J. D., Howard, A. J. & Kendall, K. 1981*a* *Nature, Lond.* **289**, 388–390; **292**, 89–90.
Birchall, J. D., Howard, A. J. & Kendall, K. 1981*b* *European Patent Publication*, no. 0021682.
Birchall, J. D., Howard, A. J. & Kendall, K. 1981*c* *European Patent Publication*, no. 0038126.
Birchall, J. D., Howard, A. J. & Kendall, K. 1981*d* *European Patent Publication*, no. 0030408.
Birchall, J. D., Howard, A. J. & Kendall, K. 1982*a* *Proc. Br. Ceram. Soc.* **32**, 25–32.
Birchall, J. D., Howard, A. J., Kendall, K. & Raistrick, J. H. 1982*b* *European Patent Publication*, no. 0055035.
Bolton, M. 1979 *A guide to soil mechanics*, ch. 4. London: Macmillan.
Brown, J. H. 1972 *Mag. Concr. Res.* **24**, 185–196.
Diamond, S. 1973 In *Pore Structure and properties of materials*, Proc. Int. Symp. RILEM/IUPAC, pp. B73–B78. Prague: Academia.
Double, D. D., Hellawell, A. & Perry, S. J. 1978 *Proc. R. Soc. Lond.* A **359**, 435–451.
Downing, S. B., Osmond, D. W. J., Skinner, M. W., West, E. J., Dawson, D. G. 1976 *U.K. Patent*, no. 1563190.
Feret, R. 1897 *Bull. Soc. Encour. Ind. natn., Paris* II, 1604.
Griffith, A. A. 1920 *Phil. Trans. R. Soc. Lond.* A **221**, 163–198.
Groves, G. W. 1981 *J. Mater. Sci.* **16**, 1063–1070.
Hansen, T. C. 1965 *J. Am. Concr. Inst.* **62**, 193–215.
Hattori, K. 1978 In *Superplasticizers in concrete*, vol. 1 (ed. V. M. Malhotra, E. E. Berry & T. A. Wheat), *Proc. Int. Symp. Ottawa, Canada, 29–31 May 1978*, pp. 49–86.
Higgins, D. D. & Bailey, J. E. 1976 *J. Mater. Sci.* **11**, 1995–2003.
Helmuth, R. A. 1980 In *7th Int. Congr. on Chemistry of Cement*, vol. III, pp. VI–0/16 to VI–030. Paris: Éditions Septima.
Helmuth, R. A. & Turk, D. A. 1966 In *Symp. on Struct. of Portland Cement Paste and Concrete*, no. 90, 135–144. Washington: Highway Res. Board.
Hill, R. 1950 *Plasticity*, pp. 226–231. Oxford University Press.
Hillemeier, B. & Hilsdorf, H. K. 1977 *Cem. Concr. Res.* **7**, 523–536.
Kaplan, M. F. 1961 *J. Am. Concr. Inst.* **58**, 591–610.
Kendall, K. 1980 *Contemp. Phys.* **21**, (3), 277–297.

Kendall, K. 1982 *Int. J. Adhesion and Adhesives*, July, pp. 149–154.
Kesler, C. E., Naus, D. J. & Lott, J. L. 1972 In *Mechanical behaviour of materials*, vol. IV, pp. 113–124. Japan: Society of Materials Science.
Lawn, B. R. & Wilshaw, T. R. 1975 *Fracture of brittle solids*, ch. 1. Cambridge University Press.
Lawrence, C. D. 1969 *Res. Notes Cem. Concr. Ass.* no. 19, 1–21.
L'Hermite, R. & Valenta, M. 1937 *Anals Inst. tech. Bâtim.* **6**, 23–27.
Lucretius 1951 *On the nature of the Universe* (transl. by R. Latham), p. 43. London: Penguin.
Mindess, S. 1981 *Materials Research Series Report*, no. 2, 93 pages. Canada: University of British Columbia.
Newman, K. 1966 In *Composite materials* (ed. L. Holliday), ch. 8. London: Elsevier.
Powers, T. C. 1960 *Proc. 4th Int. Symp. Chemistry of Cement.* Nat. Bur. Stds. Monograph, no. 43, vol. II, pp. 577–613. Washington, DC.
Rice, R. W., Freiman, S. W., Pohanka, R. C., Mecholsky, J. J. & Wu, C. C. 1978 In *Fracture mechanics of ceramics*, vol. 4, pp. 849–876. London: Plenum.
Roy, D. M. & Gouda, G. R. 1975 *Cem. Concr. Res.* **5**, 153–162.
Ryshkewitch, E. 1953 *J. Am. ceram. Soc.* **36**, 65–68.
Schiller, K. K. 1958 In *Mechanical properties of non-metallic brittle materials* (ed. H. Walton), pp. 35–49. London: Butterworth.
Swamy, R. N. 1979 *J. Mater. Sci.* **14**, 1521–1553.
Swartz, S. E., Hu, K. K. & Jones, G. L. 1978 *J. Engng. Mech. Div. Am. Soc. civ. Engs.* **104**, 789–800.
Vitruvius 1960 *The ten books of architecture* (transl. by M. H. Morgan), p. 43. New York: Dover.
Winslow, D. N. & Diamond, S. 1970 *J. Mater.* **5**, 564–585.
Wright, W. & Byrne, J. G. 1964 *Nature, Lond.*, **203**, 1374–1375.

Discussion

P. L. Pratt (*Department of Metallurgy and Materials Science, Imperial College, Prince Consort Road, London SW7 2BP, U.K.*). In their interesting series of experiments Kendall and his co-authors have shown that porous silica glass obeys Griffith's (1920) criterion for a range of porosities from 0 to 0.5. As a result of these experiments they claim to have discovered a new theory of fracture, which they express in equation (7). My first question is what is new about this theory? Surely it is a simple statement of the Griffith criterion in which particular empirical relations are used to approximate the values of E and R. As such it has been used by those interested in the fracture mechanics of ceramics for many years, although the choice of Balshin's (1949, 1950) form of the dependence of E on porosity might not be the best.

The authors claim further that the cause of the natural flaws in hardened cement paste are pores arising from air in the cement, owing to flocculation. Trapped air is present in hand-mixed cement paste and the easiest way to remove it is with a short vacuum treatment of the fresh paste. This removes the bubbles of air and increases the initial ultrasonic pulse velocity from that for transmission through air to that for transmission through water. De-airing eliminates the large pores from the microstructure but unfortunately has only a small effect on the strength and fracture toughness of well mixed hardened paste. A more likely explanation of the low tensile strength of ordinary cement paste lies in the morphology of the hydration products and their susceptibility to subcritical crack growth. Inherent flaws like inner product detached from Hadley shells (Pratt & Ghose, this symposium) are not affected by vacuum de-airing. Have the authors considered the effects of the vigorous working and limited hydration involved in the MDF process on other defects in the microstructure, like Hadley grains, and on subcritical crack growth?

References

Balshin, M. Y. 1949 *Dokl. Akad. Nauk SSSR* **67**, 831.
Balshin, M. Y. 1950 *Chem. Abstr.* **44**, 2425f.

K. Kendall. We do not claim that equation (7) is new. However, we have not been able to find any reference to this equation and its experimental verification in the literature relating either to cement or to ceramics. Indeed, we have been unable to find any statement that there is a conflict between the Ryshkewitch (1953) equation and the Griffith (1920) theory. Can Professor Pratt provide some evidence of such previous work?

On the further question of large pores in cement, air is not the only source of such pores. The other major source is packing defects, that is regions of the cement mix containing water but insufficient cement to fill the space with hydrate.

Moreover, Professor Pratt is wrong in stating that removal of air bubbles has only a small effect on the strength of hardened cement paste. Removal of air has a large effect on strength, as Feret pointed out in 1897. A 20 % volume fraction of air reduces strength by a factor 3.

B. A. Proctor (*Fibres and Composites Department, Pilkington Brothers p.l.c., Research and Development Laboratories, Hall Lane, Lathom, Ormskirk, Lancashire, U.K.*). The apparent anomaly between the effects of pores and cracks may be explained by the approach of Inglis (1913), who analysed the stress-concentrating effect of elliptical voids, in contrast with the approach of Griffith (1920), who used the stress distributions of Inglis to analyse the energetics of propagation of infinitely sharp elliptical voids or cracks.

In the work of Inglis the sharpness of the crack tip is an important factor affecting the degree of stress concentration of a given size of crack and this was well recognized in work on the strength of brittle solids some 20 years ago. Since that time the Griffith approach concentrating on the factors governing the propagation of an already infinitely sharp crack has become more widely used, and the importance of shape and tip curvature, emphasized by Inglis, rather neglected.

Reference

Inglis, C. E. 1913 *Trans. Instn nav. Archit.* **55**, 219–230.

K. Kendall. The Inglis approach cannot possibly explain anything about the fracture of brittle materials such as hardened cement, because it implies that cracks propagate as a result of *stress*, and thus argues that the *shape* of a notch (or crack) is the only important parameter, not the *length* of the notch. This idea is false.

As Griffith demonstrated with his *energy* criterion of cracking, it is the *length* of the notch that is of prime importance for brittle materials, not the *shape*, which may be varied enormously without much effect on fracture. This result may easily be demonstrated by cutting notches of different shapes in MDF cement.

Confusion on this issue partly stems from Griffith's (1920) original paper in which he states the energy criterion, but then appears to support the opposing stress argument by stating 'rupture of an isotropic elastic material always occurs at a certain maximum tension' (i.e. stress). In fact he reverted to the stress criterion in his later paper on compressive failure (Griffith 1924).

Reference

Griffith, A. A. 1924 In *Proceedings of the first International Congress on Applied Mechanics, Delft* (ed. C. B. Biezeno & J. M. Burgers), pp. 55–63.

S. A. Jefferis (*Civil Engineering Department, King's College London, Strand, London, WC2R 2IS, U.K.*).
The results have shown the dramatic effect that pore size has on strength. One reason for this must be the contrast in elastic modulus between the empty pore and the matrix. If so, unhydrated cement particles may also behave as flaws as they may be significantly stiffer than their hydration products. Such stiff flaws will have less effect than voids but could become significant in materials with low porosity and very small pores. Is cement grain size also an important factor in controlling the properties of MDF materials?

K. Kendall. Unhydrated grains of cement are indeed stiffer than their hydration products, and therefore may behave as flaws in MDF cements. When large grains of stiff material were deliberately added to the MDF cement, the flexural strength of the material was found to be reduced.

Phil. Trans. R. Soc. Lond. A **310**, 155–166 (1983) [155]
Printed in Great Britain

Novel methods of processing cement gel to examine and control microstructure and properties

By L. J. Parrott
Cement and Concrete Association, Wexham Springs, Slough SL3 6PL, U.K.

The microstructure of the gel produced by hydrating cement is sensitive to the manner in which the original pore water is removed. Exchange of the pore water with a miscible, low surface-tension fluid by a process of counter-diffusion is examined. Diffusion rates in the hydrated cement can be assessed from the kinetics of the exchange process. Evaporation of the low surface-tension fluid from the pores caused less alteration to the microstructure than direct removal of the original pore water.

The sensitivity of the microstructure of hydrated cement to normal environmental drying is industrially significant because it is associated with increases in diffusion rates that could affect the durability of concrete. Drying lowered the correlation between diffusion rate and compressive strength. Possible methods of processing the cement gel to make it more resistant to the disruptive effects of drying are discussed.

Introduction

The particular method of processing cement gel that will be a major theme in this paper is the exchange of the original pore water with a miscible fluid by a process of counter-diffusion. Various observations that have emerged during the study of this fluid exchange process will be discussed and particular emphasis will be placed upon the importance of drying history. The rate of fluid exchange gives an indication of the diffusivity of hydrated cement and the use of a low surface-tension fluid minimizes capillary tension stresses during subsequent drying. This latter point is significant when preparing samples for pore structure analysis. Fluid exchange can be used to introduce probe molecules into the pores of the hydrated cement so that they fully replace the original pore water.

The effects of drying are industrially important as well as being significant when preparing samples for porosity tests. Even very moderate drying can increase diffusion rates irrecoverably by a factor of two or three. Such changes in diffusion characteristics are readily detected by measurement of the kinetics of fluid exchange. The surface layer of concrete that is responsible for protecting the reinforcing steel in most concrete structures is subjected to environmental drying and its diffusion characteristics will therefore be modified. Furthermore the changes in microstructure induced by drying are associated with the development of irrecoverable shrinkage in the hydrated cement. This may cause stresses to develop in structural concrete, which could be of significance in design calculations.

The small increase in compressive strength that accompanies drying is largely reversible with rewetting. Since the changes in diffusion rate are not recoverable with rewetting, normal environmental exposure will lower the correlation between compressive strength and diffusion characteristics. This calls into question the prevalent view that the durability of concrete is directly related to its compressive strength.

Another aspect of the processing of cement gel is embodied in the problem as to how cement gel can be processed to make it more resistant to the effects of drying. Variables that can be manipulated on an industrial level will be considered and will include cement gel composition, the use of reactive siliceous materials, admixtures, water/cement (w/c) ratio, age and heat treatment.

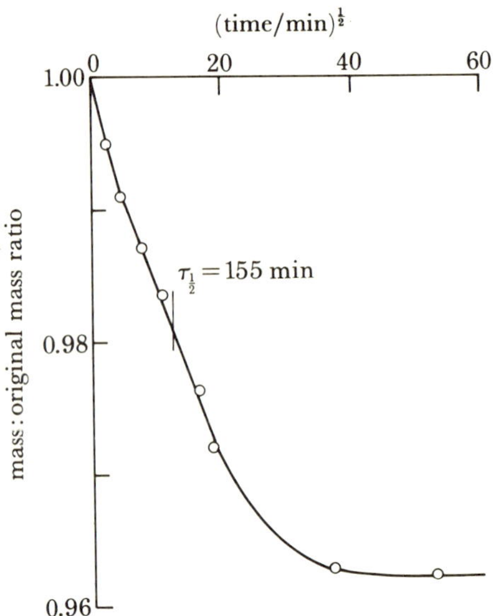

FIGURE 1. Solvent exchange for porous glass where original pore water is exchanged for propan-1-ol.

TABLE 1. POROSITY† ESTIMATES FOR A TRICALCIUM SILICATE CEMENT HYDRATED FOR 9 DAYS

sample	fluid	porosity	fluid	porosity	fluid	porosity	fluid	porosity
1	water	0.493	methanol	0.471	*n*-pentane	0.433	benzene	0.449
2	water	0.515	ethanol	0.494	*n*-pentane	0.468	benzene	0.472
3	water	0.502	propan-1-ol	0.487	*n*-pentane	0.446	benzene	0.460
4	water	0.497	propan-2-ol	0.486	*n*-pentane	0.457	benzene	0.463

† Fraction of total volume.

FLUID EXCHANGE

Fluid exchange typically involves the immersion of a water-saturated sample of porous material in a water-miscible fluid such as methanol or propanol and measuring the weight change of the sample as a function of time. Normally, 3 mm thick slabs of mass *ca.* 1 g were immersed in 150 ml of exchange fluid. Dependant upon sample pore structure, mass equilibrium was reached after periods ranging from a few hours to a few weeks. Figure 1 illustrates a solvent exchange curve for propan-1-ol diffusing into a porous glass that has 4 nm diameter pores and about 30 % porosity. Secondary and tertiary exchanges were feasible with fluids that did not need to be miscible with water. The counter-diffusion process generally followed a standard three-dimensional diffusion equation, the diffusion rate being linearly related to the viscosity of the penetrating fluid (Parrott 1982).

Table 1 illustrates the porosity data obtained from a tricalcium silicate cement. The porosity estimates were based on the various fluid-saturated weights, subtraction of a final dry mass and division by the measured density of the exchange fluid (Parrott 1982). The variations between samples for a single fluid suggest that porosity differences of about 0.02 can be reliably detected. The estimates of porosity based on water penetration are slightly higher than those determined with organic fluids. However, it is important to note that pore water replacement levels for fluid exchange are much higher than those obtained after direct drying and adsorption (Mikhail 1966).

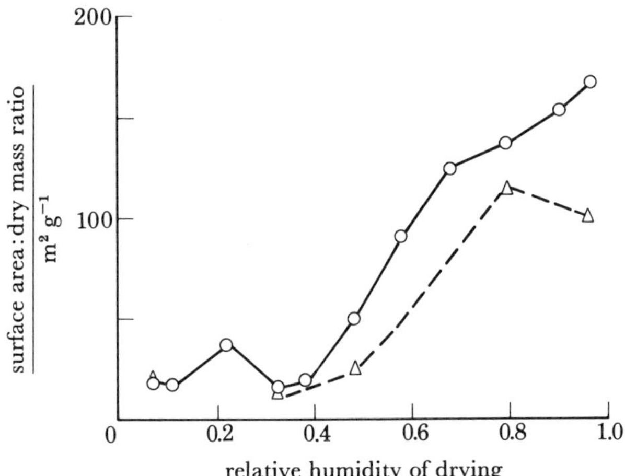

FIGURE 2. Effect of previous drying for 78 days at different relative humidities upon the nitrogen surface area of solvent exchanged and dried tricalcium silicate hydrate. *Key*: ○, w/c ratio 0.6; △, w/c ratio 0.4.

Evidence that solvent exchange followed by evaporation of the solvent produces less pore structure alteration, than does direct drying of pore water, is available from tests on several materials. Swanson (1979) demonstrated that the surface area measured by nitrogen adsorption on samples of wood pulp increased from 0.6 to 140 $m^2\ g^{-1}$ with a reduction in the surface tension of the exchange fluid from 0.72 to 0.15 $mN\ cm^{-1}$. Iler (1979) has reviewed comparable studies on the drying of silica gels from low surface tension fluids and concluded that solvent replacement was an effective way of minimizing pore shrinkage and loss of surface area. The effect of previous drying upon the surface area of two hydrated tricalcium silicate cements is illustrated in figure 2 (Parrott *et al.* 1980). The surface area was measured by nitrogen adsorption after fluid exchange and drying. The results indicate that calcium silicate hydrate, the main cementing component in hydrated Portland cement, is very sensitive to the capillary tension stresses developed during conditioning at the different relative humidities. Data comparable to those shown in figure 2 have been obtained for Portland cement by using nitrogen adsorption (Hunt *et al.* 1960; Litvan 1976), low angle X-ray scattering (Winslow 1973) and calorimetry (Sellevold & Bager 1980). Sellevold & Bager used low temperature, scanning calorimetry to measure the freezing point of pore water; the depression of freezing point is an indication of pore radius. Sellevold & Bager's results in figure 3 show trends similar to those in figure 2.

Additional evidence that solvent exchange followed by drying minimizes the stresses that act upon the cement hydrates can be derived from measurements of dimensional changes. The exchange of pore water for methanol in hydrated tricalcium silicate cements and secondary exchange of the methanol for *n*-pentane did not cause any significant dimensional change (i.e. strain changes were less than 3×10^{-5}, Parrott & Young (1982)). Furthermore, irrecoverable shrinkage strains due to methanol desorption and adsorption were about 40×10^{-5} compared with 450×10^{-5} for desorption and adsorption of water (Parrott 1981).

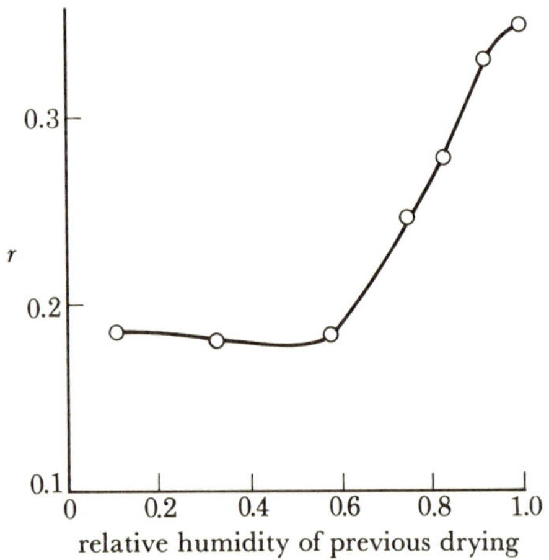

FIGURE 3. Effect of previous drying for several months upon the volume of small pores in a mature Portland cement paste (w/c ratio = 0.6 by mass) as defined by the mass ratio, r, of pore water not frozen at -10 °C to dry mass (Sellevold & Bager 1980).

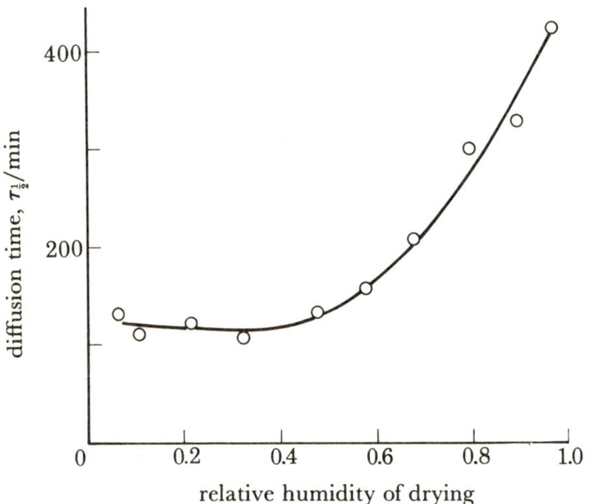

FIGURE 4. Effect of previous drying for 78 days at different humidities followed by resaturation upon the rate of exchange of pore water with methanol (water/tricalcium silicate ratio 0.6 by mass).

SIGNIFICANCE OF CHANGES IN MICROSTRUCTURE DUE TO DRYING

The effect of drying followed by resaturation upon the diffusion rate in a hydrated tricalcium silicate cement is illustrated in figure 4, where the time for half of the pore water to be exchanged with methanol, $\tau_{\frac{1}{2}}$, is plotted against the relative humidity during previous drying. The diffusion rate was increased by a factor in excess of three by prolonged drying at relative humidities less than 60% (Parrott 1981). Subsequent methanol adsorption measurements on the diffusion test samples showed that the minimum in $\tau_{\frac{1}{2}}$ corresponded to a maximum in the volume of pores larger than 5 nm and a minimum in the volume of smaller pores. Drying caused a coarsening of the structure without any great change in total porosity. This is in general agreement with the views of Sellevold & Bager (1980). Related studies by Powers *et al.* (1954) demonstrated that prolonged drying of a well hydrated Portland cement to 79% relative humidity caused a

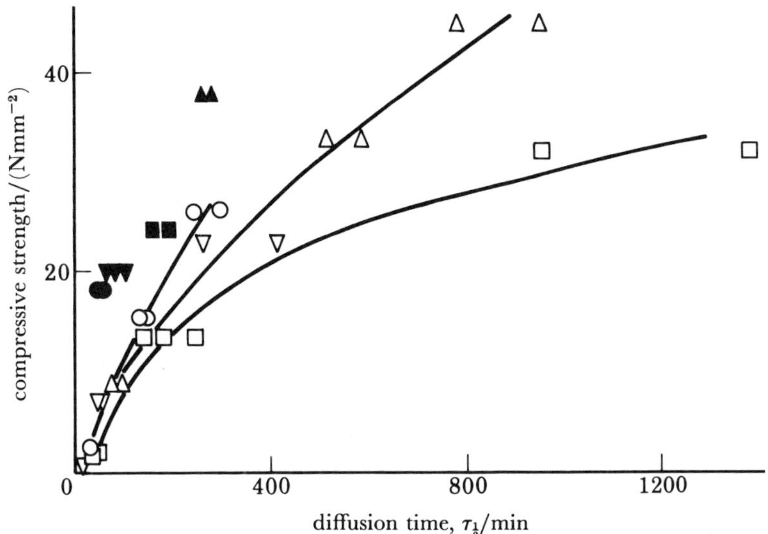

FIGURE 5. Diffusion time, $\tau_{\frac{1}{2}}$, against compressive strength for a variety of hydrated cements (w/c ratio 0.59 by mass). Cement: ○, ●, tricalcium silicate; □, ■, 0.75 tricalcium silicate with 0.25 dicalcium silicate; ▽, ▼, 0.50 tricalcium silicate with 0.50 dicalcium silicate; △, ▲, ordinary Portland cement. Open symbols, samples were not dried; filled symbols, samples were dried and resaturated.

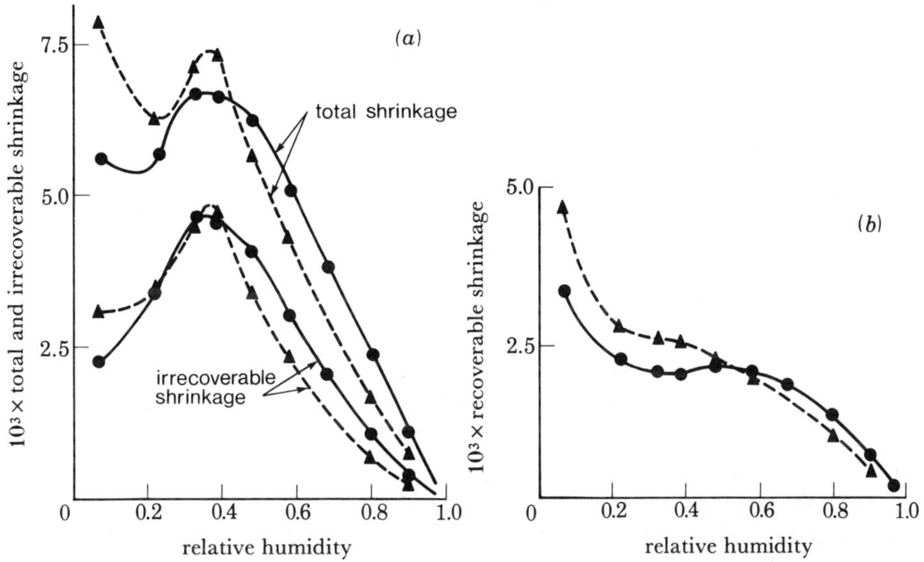

FIGURE 6. Shrinkage of hydrated tricalcium silicate (alite) cement against relative humidity of drying. Dried for 78 days, resaturated for 21 days. *Key*: ●, water/alite ratio 0.6; ▲, water/alite ratio 0.4.

70-fold increase in permeability to water. Both Powers and Sellevold commented that the primary effect of drying was to increase the continuity of the larger pores.

Drying can cause an increase in the compressive strength of hydrated Portland cement (Parrott 1973) but this effect seems to be largely recoverable upon resaturation (Parrott 1977). The effect of drying upon diffusion rate that was mentioned previously was irrecoverable and it follows that correlation between diffusion rates and compressive strength will be reduced by including the effects of drying. This is illustrated in figure 5 where the diffusion time, $\tau_{\frac{1}{2}}$, as measured by methanol exchange, is plotted against compressive strength for a variety of cements.

Consideration of the data from dried and resaturated samples shows that scatter increased to such an extent that compressive strength cannot provide a reasonable indication of diffusion characteristics. Thus the engineering tradition of equating the quality and durability of concrete to a standard measure of compressive strength requires close examination and some revision.

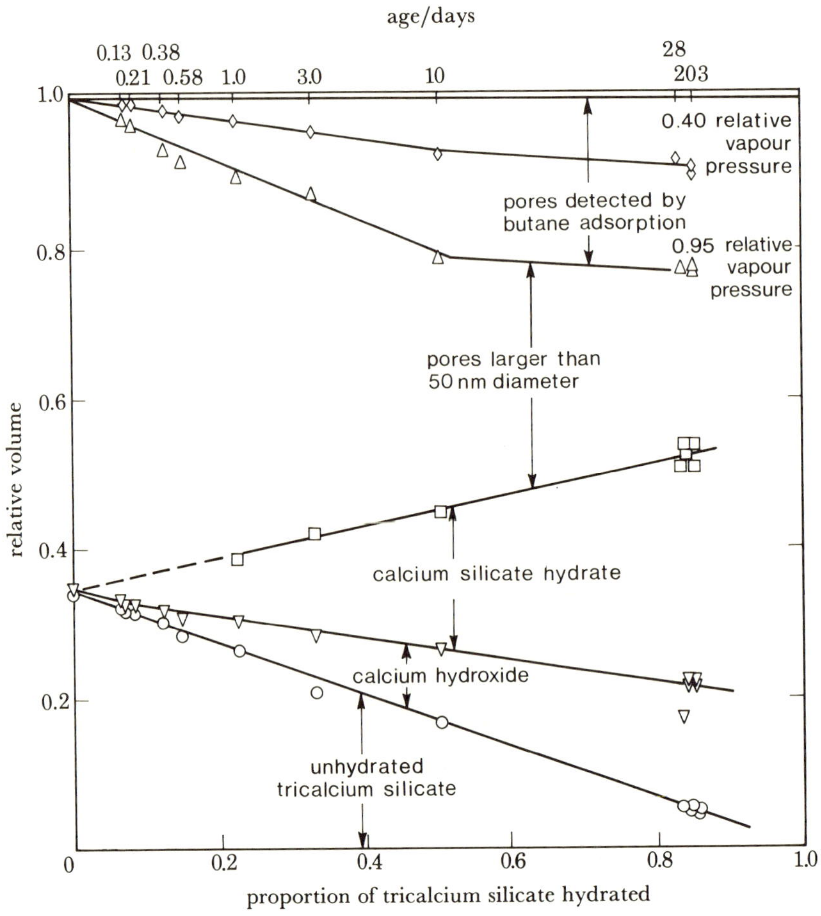

FIGURE 7. Phases present in a simple tricalcium silicate cement as a function of hydration (w/c ratio 0.59 by mass).

Shrinkage of hydrated cement is a direct consequence of drying and, as illustrated in figure 6, it is only partly recoverable with resaturation (Parrott & Young 1982). The pattern of irrecoverable shrinkage against relative humidity of previous drying has obvious parallels with the diffusion results shown in figure 4 and indeed the explanations are similar. Experiments involving methanol exchange followed by drying and methanol adsorption indicated that irrecoverable shrinkage is a direct reflexion of the internal stresses that cause a reduction in volume of the smaller pores and a corresponding increase in the volume of larger pores. The shrinkage results suggested that the stresses originated from capillary tension forces (Parrott & Young 1982).

Processing cement gel to resist the effects of drying

The earlier sections of this paper have shown how the coarsening of porosity in hydrated cement that results from drying can be controlled in the laboratory by the use of fluid exchange followed by fluid evaporation. This approach can be used in the industrial production of silica gels (Iler 1979) but it leaves the gel susceptible to capillary tension stresses if it is subsequently exposed to water vapour (Johnson & Ries 1950). Clearly a different approach is needed for industrial production of cement-based products that remain impermeable and have good dimensional stability under fluctuating water vapour pressures.

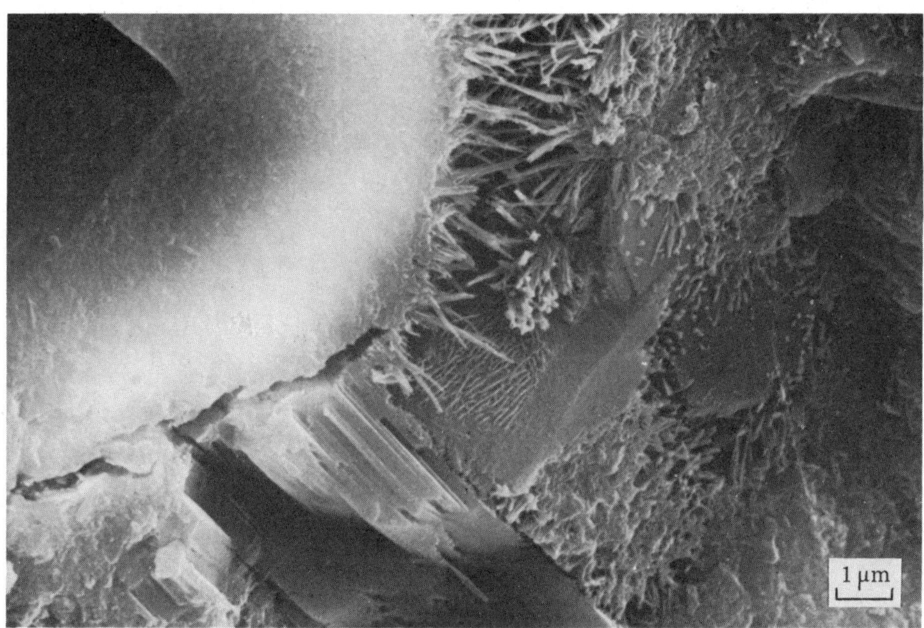

FIGURE 8. Scanning electron micrograph of fracture surface in a hydrated tricalcium silicate cement hydrated for 200 days (w/c ratio 0.59 by mass).

It is helpful, when considering how cement gel can be processed to resist the effects of drying, to review the phase volume composition of a simple hydrated tricalcium silicate cement. The phase volume diagram shown in figure 7 was based upon adsorption studies to provide porosity data, and thermogravimetric analysis to give volumes of calcium hydroxide and unhydrated alite (an impure form of tricalcium silicate cement). The spacial relations of these phases can be assessed in a qualitative manner by electron microscopy. The calcium silicate hydrates seem to coat the hydrating alite grain and the outer region is composed of radially oriented acicular growths. This is illustrated in figure 8, where the more compact, featureless inner hydrate can be seen adjacent to the angular particle of unhydrated alite. The calcium hydroxide is able to grow as fairly large crystals several micrometres in size. It is evident from scanning transmission electron microscopy that the acicular growths from adjacent cement grains intergrow and provide some skeletal continuity (Jennings *et al.* 1981).

The effect of ageing is, given sufficient water, to increase cement hydration and reduce the volume of the pores originally filled by the mixing water with a porous calcium silicate hydrate gel and calcium hydroxide, as illustrated in figure 7. While this reduces diffusion rates and increases strength, hydration will not greatly reduce the sensitivity of the porous cement gel to shrinkage during subsequent drying. Since drying does not greatly alter the total volume of the hydrated cement, nor does it significantly reduce the volume of the solid phases, the shrinkage of the porous gel increases the volume of the larger pores and thus increases the rate of diffusion. Reductions in w/c ratio normally reduce the volume of larger pores without greatly affecting the other phases, and the material remains susceptible to the effects of drying. This view is supported by comparison of the changes in diffusion times, $\tau_{\frac{1}{2}}$, resulting from drying for two hydrated tricalcium silicate cements of 0.6 and 0.4 w/c ratio, table 2.

Table 2. Effect of drying for 78 days upon diffusion rates in two hydrated calcium silicate cements (Parrott 1981)

	diffusion time, $\tau_{\frac{1}{2}}$/min		
w/c ratio	not dried	dried 80% r.h.	dried 40% r.h.
0.4	1850	1100	250
	(1.00)	(0.59)	(0.14)
0.6	400	180	75
	(1.00)	(0.45)	(0.19)

The changes relative to the undried samples show that lowering the w/c ratio did not significantly reduce sensitivity to drying.

Tricalcium silicate and dicalcium silicate are major constituents of Portland cements and they hydrate to form similar calcium silicate hydrates with different quantities of calcium hydroxide (Kantro *et al.* 1966). Since the morphology of calcium silicate hydrate differs from that of calcium hydroxide (Diamond 1976 and figure 8) the relative quantities of tricalcium silicate and dicalcium silicate in a cement could, in theory, affect diffusivity. Diffusion measurements in which pore water was exchanged for methanol were obtained for cements blended from tricalcium silicate and dicalcium silicate. The results are plotted in figure 5 against compressive strength and it can be observed that for a given strength a wide range of $\tau_{\frac{1}{2}}$ values were obtained with non-dried samples. An intermediate proportion of dicalcium silicate (25% of the total weight of cement) gave the longest diffusion times, suggesting that the quantity of calcium hydroxide was not a dominant factor. A more fundamental view of the diffusion results is illustrated in figure 9 where $\tau_{\frac{1}{2}}$ is plotted against the volume of pores larger than 4 nm diameter. Scatter is smaller than that in figure 5 and the dried and non-dried samples conform to a similar relation. Figure 9 suggests that diffusion is dominated by the larger pores and the relative quantities of tricalcium silicate and dicalcium silicate have only an indirect effect on diffusion via control of the rate of hydration and pore filling. The increase in tricalcium silicate: dicalcium silicate ratio from 1:1 to 3:1 corresponds roughly to the change in the relative proportions of these minerals in ordinary Portland cement over the last sixty years (Corish & Jackson 1982). Figures 5 and 9 suggest that this change is beneficial for non-dried cements but a cautionary note should be added that if the moist conditions encourage dissolution of the calcium hydroxide the benefit might be lost. In drying conditions any effect of tricalcium silicate to dicalcium silicate ratio in the cement upon diffusion characteristics would be masked by the change in pore structure associated with drying (figure 5).

Preliminary studies with cements containing calcium aluminates suggest that the porosity–diffusion relation shown in figure 9 is not unique and at early ages it is possible to obtain longer diffusion times for a given volume of large pores. The reasons for this are currently under investigation.

Materials that contain reactive silica such as pulverized fuel ash and silica fume can be blended with Portland cement so that their reaction with the calcium hydroxide from the cement reaction yields additional calcium silicate hydrates. Silica fume reacts with calcium hydroxide to yield a calcium silicate hydrate that is not significantly greater in volume than the reactants (Sellevold *et al.* 1981). However, the calcium silicate hydrate forms as a porous gel and the gel pores contribute to the filling of the larger pores. Thus the addition of silica fume to a Portland cement reduced the volume of larger pores and reduced diffusion rates without any reduction in total porosity. Nevertheless the porous gel was still sensitive to drying, with corresponding increases in the volume of larger pores.

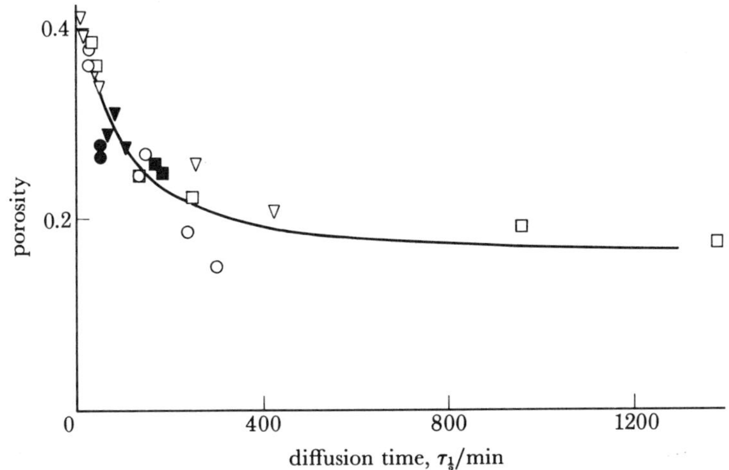

FIGURE 9. Diffusion time, $\tau_{\frac{1}{2}}$ against volume of pores larger than 4 nm diameter (w/c ratio 0.59 by mass). Cement: ○, ●, tricalcium silicate; □, ■, 0.75 tricalcium silicate with 0.25 dicalcium silicate; ▼, ▽, 0.50 tricalcium silicate with 0.50 dicalcium silicate. Open symbols, samples were not dried; filled symbols, samples were dried and resaturated.

Sereda *et al.* (1980) have reviewed the effect of admixtures upon the microstructure of hydrated cement. There was no evidence to suggest that any commonly used admixtures can limit the increases in diffusion rate that result from first-time drying. Where admixtures affected cement hydrate morphology and pore structure they seemed to leave the hydrated cement more susceptible to the disruptive effects of drying. This was deduced from the development of irrecoverable shrinkage (Feldman & Swenson 1975).

Heat treatment of saturated hydrated cement has been used to stabilize it against changes due to subsequent drying, but unfortunately heating leads to a coarsening of the pore structure similar to that induced by drying (Sellevold & Bager 1980). This coarsening of the pore structure was also observed in a study of the effects of heating upon creep and shrinkage in hydrated Portland cement (Parrott 1977). Thus it seems unlikely that heat treatment will yield a hydrated cement with reduced diffusion rates.

The effects of drying upon the pore structure and diffusion properties of hydrated cement are substantial and are of industrial importance, particularly with regard to the durability of concrete. At present there appears to be no obvious way of making hydrated cement resistant to the disruptive effects of drying and further work on this topic seems desirable.

Conclusions

Hydrated cement can be processed by using fluid exchange techniques in which pore water is exchanged for another miscible fluid by a process of counter-diffusion. The kinetics of the exchange process are indicative of the diffusion properties of the hydrated cement. If the exchange fluid has a low surface tension then capillary tension stresses during subsequent drying are reduced. Correspondingly, shrinkage and pore structure changes are reduced and later pore structure analysis is thereby facilitated.

Under practical conditions the drying of hydrated cement does not involve fluid exchange and substantial coarsening of the pore structure occurs. This coarsening leads to faster diffusion rates and is therefore industrially important particularly with regard to the durability of concrete. Drying lowered the degree of correlation between compressive strength and diffusion rate, indicating that compressive strength is not a reliable measure of durability.

Further work is needed to produce a hydrated cement that is resistant to the disruptive effects of drying.

The author wishes to thank Miss K. Scrivener of the Department of Metallurgy and Materials Science, Imperial College, London for providing the micrograph in figure 8.

References

Corish, A. T. & Jackson, P. J. 1982 *Concrete* **16**, 16–18.

Diamond, S. 1976 In *Proc. Conf. Hydraulic Cement Pastes; their Structure and Properties, Sheffield*, pp. 2–30. Slough: Cement and Concrete Association.

Feldman, R. F. & Swenson, E. G. 1975 *Cem. Concr. Res.* **5**, 25–35.

Hunt, C. M., Tomes, L. A. & Blaine, R. L. 1960 *J. Res. nat. Bur. Stand.* A **64** (2), 163–169.

Iler, R. K. 1979 In *The chemistry of silica*, pp. 533–539. New York, Chichester, Brisbane and Toronto: John Wiley and Sons.

Jennings, H. M., Dalgleish, B. J. & Pratt, P. L. 1981 *J. Am. ceram. Soc.* **64**, 567–572.

Johnson, M. F. L. & Ries, H. E. 1950 *J. Am. chem. Soc.* **72**, 4289.

Kantro, D. L., Weise, C. H. & Braunauer, S. 1966 *Highw. Res. Abstr. report* no. 90, 309–327.

Litvan, G. G. 1976 *Cem. Concr. Res.* **6**, 139–144.

Mikhail, R. Sh. 1966 *Highw. Res. Abstr. report* no. 90, 123–134.

Parrott, L. J. 1973 *Mag. Concr. Res.* **25**, 197–200.

Parrott, L. J. 1977 *Chem. Concr. Res.* **7**, 597–604.

Parrott, L. J. 1981 *Chem. Concr. Res.* **11**, 651–658.

Parrott, L. J. 1982 Examination of two methods for studying diffusion kinetics in hydrated cements. Materials and Structures. (In the press.)

Parrott, L. J., Hansen, W. & Berger, R. L. 1980 *Chem. Concr. Res.* **10**, 647–655.

Parrott, L. J. & Young, J. F. 1982 In *Proc. Conf. Fundamental Res. on Creep and Shrinkage of Concrete, Lausanne* (ed. F. H. Wittmann), pp. 35–48. The Hague, Boston and London: Martinus Nijhoff Publishers.

Powers, T. C., Copeland, L. E., Hayes, J. C. & Mann, H. M. 1954 *J. Am. Concr. Inst.* **26**, 285–298.

Sellevold, E. S., Bager, D. H., Jensen, E. K. & Knudsen, T. 1981 *Proc. Symp. Silica in Concrete.* Trondheim: Cement and Concrete Research Institute.

Sellevold, E. J. & Bager, D. H. 1980 *Technical University of Denmark*, *Technical Report* no. 86/80.

Sereda, P. J., Feldman, R. F. & Ramachandran, V. S. 1980 In *Proc. 7th Int. Symp. Chemistry of Cement*, **1**, subtheme VI-1.

Swanson, J. W. 1979 In *Proc. Symp. Characterisation of Porous Solids, Neuchatel* (ed. S. J. Gregg, K. S. W. Sing & H. F. Stoeckli), pp. 339–350. London: Society of Chemical Industry.

Winslow, D. N. 1973 Ph.D. thesis, Purdue University.

Discussion

K. S. W. Sing (*Department of Chemistry, Brunel University, Uxbridge, U.K.*). I wish to make two comments concerning the interesting results presented by Dr Parrott.

(1) The solvent exchange technique is similar to the procedure introduced many years ago by Kistler to prevent the collapse of pore structure that normally takes place during the drying of an aerogel (i.e. conversion to the more compact xerogel). Various additives have been used to modify the surface structure of alumina and silica to render the surface more hydrophobic and it is possible that this type of approach could be used as a means of stabilizing the pore structure of the cement hydrogel.

(2) It should be noted that the pore structure of the cement gel is not rigid even when the continuous aqueous phase has been removed. Special care is required therefore in the interpretation of adsorption and other data in the assessment of pore size distribution.

L. J. Parrott. With regard to Professor Sing's first point I would like to thank him for his suggestion and I will examine the use of additives that create a hydrophobic surface for stabilizing the pore structure of cement gel.

I fully agree with the second point in which Professor Sing states that the gel may not be stable even when the pore water has been removed. Fortunately this did not seem to be a major problem in the present work. As I mentioned in my paper, results from adsorption experiments showed the same trends as results from the low temperature calorimetry where drying of the cement was not necessary.

A. A. Rahman (*Aberdeen University, U.K.*). Results presented by Dr Parrott on the changes of surface area, amounts of water not frozen at -10 °C and diffusion rates with relative humidity showed interesting features. There was a maximum for surface areas (nitrogen) at around 0.2 r.h., while the trends (slopes of the lines) for unfrozen water and diffusion rates changed at about 0.5 r.h. Is there an explanation for these? Could these be explained in terms of removal of water or other adsorbed molecules?

The surface areas reported would be affected by the nature of interaction between the gel surface and the adsorptive, particularly the specificity component of the interaction. The displacement of one fluid by another may leave a residual adsorbed layer of the initial (liquid) molecules. Is there any evidence for the specificity of interaction with the gel system and the several molecular probes Dr Parrott used? Further, how has the complete removal of the adsorptive layer been ascertained? Is it likely that the surface areas reported would represent values for a gel-adsorptive surface rather than a true gel surface?

L. J. Parrott. Nitrogen surface areas, water not frozen at -10 °C, diffusion times and published methanol adsorption results (Parrott 1981) all show a major decline with drying in the relative humidity range from saturation down to 0.5. The most obvious explanation for this behaviour is that surface tension forces cause compaction of the porous cement gel and thereby alter the pore size distribution.

In reply to Dr Rahman's second point there was remarkably little evidence of an interaction between the cement gel and probe molecules. This seemed to be related to the method of introducing the probe molecule; normal drying compaction of the porous gel can produce molecular sieve effects that might be regarded as an interaction between gel and probe molecule but with fluid exchange the gel remains dispersed and non-polar fluids with relatively large molecular radii can replace a comparable volume of pore water (see table 1).

D. Pearson (*A.E.R.E. Harwell, U.K.*). Recent work using small angle neutron scattering has indicated results that are in close agreement with those presented by Dr Parrott. From looking at pore size distributions below about 25 nm, it appears that drying at either 105 °C or 11% r.h. does have a significant effect on pore structure. Changes in water/cement ratio or additions of silica fume, on the other hand, appear to have little effect on gel porosity.

Phil. Trans. R. Lond. A **310**, 167–173 (1983)
Printed in Great Britain

Development and application of high-density cement-based materials

By L. Hjorth

Aalborg Portland, P.O. Box 165, *Rørdalsvej* 44 *DK*-9100 *Aalborg, Denmark*

The properties of cement-based binders change dramatically when ultrafine particles are homogeneously placed in the spaces between densely packed cement grains. The increased density and the refinement of the pore structure of such binders result in considerable increases in properties such as compressive strength, impermeability and adherence to fibres and aggregates. It has been shown that such cement-based binders may be processed by simple techniques, and that the resulting materials may replace metals, ceramics and plastics.

Introduction

In the cement and concrete industry, the research and development organizations have often felt a challenge to increase the strength of cement-based materials. The maximum compressive strength of concrete is used as a direct measure of the development possibilities. In a few cases only, the tensile strength has been used as a similar measure although increase in tensile strength offers a wider scope for development.

As the strength increases with decreasing porosity and with smaller pore sizes (Powers *et al.* 1963, Birchall *et al.* 1981) the main principle in the development work towards increased strength has been to reduce the total porosity of the cement paste and to optimize the pore size distribution by various techniques. These techniques have included:

(i) mechanical treatments of the cement paste such as vibrational compaction (Bache *et al.* 1968), high pressure pressing (Roy *et al.* 1973; Lawrence 1969) and high shear mixing (Birchall *et al.* 1980);

(ii) the use of effective cement dispersing compounds, which reduce the water content necessary to obtain a 'workable' consistency of the cement paste (Hattori 1978);

(iii) filling up the pores with solid materials, such as sulphur (Malhotra 1975), and resin (Idorn *et al.* 1974);

(iv) the use of cements with special particle size distributions, such as very fine-grained cement (Brunauer 1972), and gap-graded cement (Birchall *et al.* 1980).

The present paper describes results from the industrial development of high-strength cement-based materials in which several of these principles are combined.

During the development work of these materials, extremely high compressive strengths were achieved, but it was also found that other properties were changed considerably, and realized that a new class of materials might be produced (Bache 1981). For the sake of brevity the new materials are termed D.S.P. materials referring to 'densified systems containing homogeneously arranged ultrafine particles'.

Properties of D.S.P. materials

D.S.P. binders may be produced from Portland cement, microsilica and superplasticizers. Superplasticizers are very effective dispersing agents often used in the concrete industry to reduce the water required to achieve a workable concrete. Microsilica is a by-product from the silicon and ferrosilicon industries. It has been used for some years to improve the properties of concrete (Hjorth 1982). Microsilica particles are spherical in shape and have an average diameter of about 0.1 μm. This size is about two orders of magnitude smaller than the cement grains, which indicates that microsilica may act as a microfiller in between the cement grains. Since microsilica is almost pure amorphous silica, it reacts chemically with the ions in the alkaline pore solution of the cement paste, forming reaction products which resemble the hydration products, found in ordinary, hardened cement paste. The D.S.P. binders consist of densely packed cement and these ultrafine microsilica particles homogeneously arranged in the space between the cement grains.

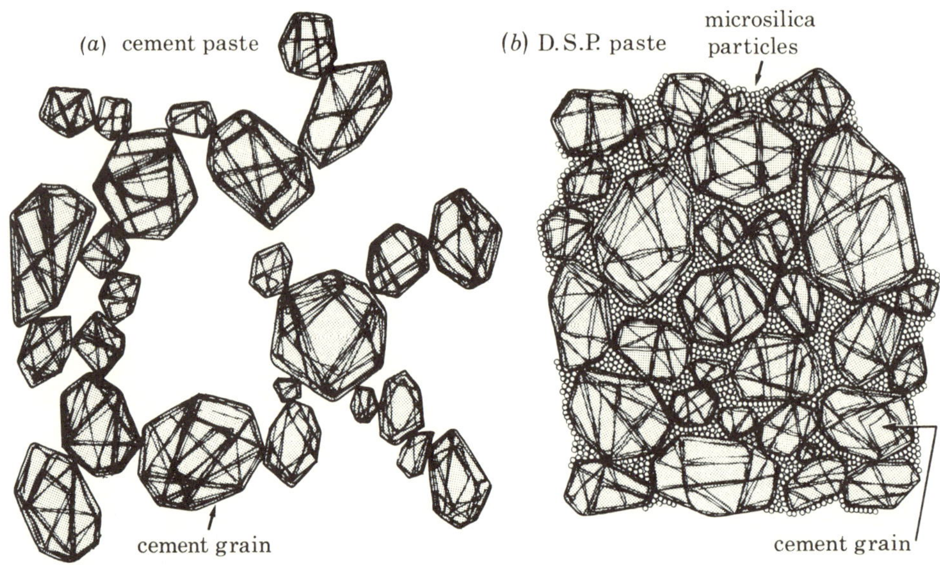

FIGURE 1. (*a*) Flocculated particles in ordinary cement paste. (*b*) Densely packed cement grains and microsilica particles in D.S.P. paste.

Production of the new materials is made possible by the superplasticizers, which change the flocculent, fine particle systems of cement and microsilica to systems that can be densely packed in a low stress field. Figure 1 (*a*) illustrates the difficulty of achieving dense packing of the cement particles, which is because the suspension of ground cement clinker plus gypsum in water is strongly flocculated (Diamond 1980). The cement grains adhere to each other, and the surface forces between the particles tend to prevent them from sliding relative to each other during mixing and casting. Figure 1 (*b*) illustrates, in contrast, the effective dispersing action of superplasticizers in deflocculating the system and enabling dense packing of the cement grains, and the microsilica particles in the voids between the cement grains.

The D.S.P. materials constitute a range of composite materials, comprising D.S.P. binders and additional solid particles such as aggregates and fibres. The properties of these materials can be varied considerably according to the composition and to the processing technique. The water to powder ratio (i.e. water to cement plus microsilica ratio) of these materials is generally within the range 0.12–0.22.

Such materials can be designed to have consistencies ranging from 'flowing' to 'plastic'. They can, therefore, be easily processed by simple techniques such as casting. More sophisticated processing techniques such as extrusion and rolling may also be applied, owing to the high internal cohesion, which is characteristic of D.S.P. materials. The strength properties are illustrated in figure 2, which shows a compressive strength of 270 MPa for a cast D.S.P. material. The tensile strength of such a product is about one tenth of the compressive strength.

Without reinforcement, the new materials are distinctly brittle with a linear stress–strain curve right up to the point of failure. The very dense microstructure of the D.S.P. binder, however, makes it particularly suitable for mechanical anchoring of thin fibres, which are not normally very firmly fixed in ordinary cement paste. The addition of such fibres gives the strong, brittle D.S.P. matrix a high degree of ductility (Bache 1981).

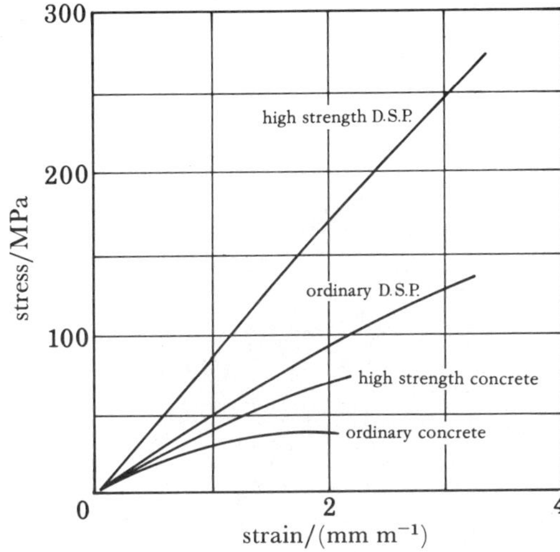

FIGURE 2. Stress–strain diagrams for ordinary concrete and D.S.P. materials.

Although the compressive strength is high, it is still only half of the strength of high quality structural steel. The density of the steel, however, is three times that of the D.S.P. materials, thus the strength:density ratio for D.S.P. materials is excitingly high. With the development of D.S.P. materials, we may have the means of building far bigger structures (bridges and towers) than has hitherto been possible.

Electron microscope observations reveal that D.S.P. materials have a very compact and dense microstructure. The addition of microsilica to cement paste is known to change the pore size distribution of such materials considerably towards finer pores (Sellevold *et al.* 1982), and a variety of observations confirm the extremely fine pore structure of D.S.P. binders.

Preece *et al.* (1982) found no freezable water in D.S.P. binders down to temperatures as low as -35 °C, and they found the Cl^- diffusion coefficient to be more than one order of magnitude lower than that of cement paste. Arup (1980) found the electrical resistivity of D.S.P. materials at 100% r.h. to be about 10^4 Ω m, which is 2–3 orders of magnitude greater than that of cement mortar, and close to the resistivity required for electric insulators.

APPLICATION OF D.S.P. MATERIALS

Although the first D.S.P. materials were made in the laboratory only five years ago, and although the properties of these materials are not yet fully established, the materials have already been used in practice for a range of purposes. Patent applications for D.S.P. materials and products are pending in several countries, and the D.S.P. materials and D.S.P. products are marketed by Densit A/S, Denmark, and Elborg Technology Company, U.S.A.

The earliest applications were overlays to concrete that was exposed to various types of aggressive environment, for example floors in chemical industries, parking areas and balconies. As liquid flow and the rate of ion diffusion are considerably reduced in these materials, deterioration by mechanisms that involve mass transfer is correspondingly decreased. Subsequently, more sophisticated applications of D.S.P. materials were attempted. In particular, the high abrasion-resistance of D.S.P. materials that use calcined bauxite as aggregates, has been utilized in materials that substitute cast basalt, rubber and steel as lining materials. Figure 3 shows a screw for feeding fly-ash into the pneumatic powder-transfer system at a cement plant. Such screws are normally made out of steel covered with sintered carbide. The steel screw is normally worn out within 250 h, whereas a similar screw, cast in D.S.P. material, had a service life that was five times longer.

FIGURE 3. Screw cast in D.S.P. material.

Cast D.S.P. material is able to reproduce a surface geometry exactly while retaining high mechanical strength. D.S.P. materials are therefore used for press tools for pressure-shaping steel plates in the car body industry. The tools are made by simple casting methods at room temperature, with the original body parts as formwork.

CONCLUSION

Owing to suitable physical properties combined with simple processing techniques and low energy consumption, the D.S.P. materials have substituted successfully for cast basalt, ceramics, polymers, cast iron, steel, brass and other metals, in a wide variety of applications. The development of the D.S.P. materials is one example among others (Birchall *et al.* 1981), which illustrates that we are far from fully utilizing the large potentials of 'today's cements'. Exploitations have started, however, and we may already consider cement as an almost unlimited source for making inorganic products by a low-temperature technology, in competition with the high-temperature solidification technology of metals and with the resource-deficient technology of organic plastics. The development further suggests that fundamental studies and application of results from colloidal science, silicate chemistry, and the chemistry of highly concentrated and highly alkaline solutions, may gradually create the cement of tomorrow, which may be the basic raw material for a wide range of products in the future, thereby instituting a 'silica-age'.

References

Arup, H. 1980 *European Patent Application* no. 81.105310.7. Assigned to Aalborg Portland.
Bache, H. H. 1981 Densified cement/ultrafine particle-based materials. Presented at the Second International Conference on Superplasticizers in Concrete, 10–12 June 1981, Ottawa, Ontario, Canada.
Bache, H. H., Lystbæk, M. & Mikkelsen, A. 1968 Komprimering af cement og finmalet kvarts ved kombineret tryk og vibrering. Orienterende undersøgelse, *B.F.L. Internal Report no. 189*, 34 pages, Betonforskningslaboratoriet, Karlstrup.
Birchall, J. D., Howard, A. J. & Kendall, K. 1981 *Nature, Lond.* **289**, 388–390.
Birchall, J. D., Kendall, K. & Howard, A. J. 1980 *European Patent Application* no. 80301909.0. Assigned to Imperial Chemical Industries p.l.c.
Brunauer, S. 1972 *U.S. Patent* no. 3689294. Assigned to the Research Corporation.
Diamond, S. 1980 *Wld. Cem. Technol.* **11**, 116–121.
Hattori, K. 1978 In *Superplasticizers in concrete* (ed. V. M. Malhotra, E. E. Berry & T. A. Wheat), pp. 49–86. Proceedings of an International Symposium held in Ottawa, Canada, 29–31 May 1978.
Hjorth, L. 1982 Microsilica in Concrete. *Nordic Concrete Research Publication no. 1* (ed. The Nordic Concrete Federation, Oslo), pp. 901–918.
Idorn, G. M., Fördös, Z. 1974 Cement–Polymer Materials. *The VI International Congress on the Chemistry of cement, Moscow*. Principal paper III. 9.
Lawrence, D. D. 1969 *Cement and Concrete Association Research Rep.* no. 19,
Malhotra, M. 1975 *J. Am. Concr. Inst.* **72**, 466–473.
Powers, T. C. & Brownyard, T. C. 1948 *Portland Cem. Ass. Res. Dept. Bull.* no. 22.
Preece, C. M., Frølund, T., Bager, D. H. 1982 In *Condensed silica fume in concrete* (ed. O. E. Gjørv & K. E. Løland), pp. 51–59. Trondheim: The University of Norway.
Roy, D. M., Gouda, R. R. 1973 *J. Am. ceram. Soc.* **56**, 549–550.
Sellevold, E. J., Bager, D. H., Klitgaard Jensen, E. & Knudsen, T. 1982 In *Condensed silica fume in concrete* (ed. O. E. Gjørv & K. E. Løland), pp. 19–31. Trondheim: The University of Norway.

Discussion

N. McN. Alford (*I.C.I., The Heath, Runcorn, Cheshire, U.K.*). I should like to ask what the fracture energy is of Dr Hjorth's material.

L. Hjorth. The D.S.P. binder as such is brittle. Fibre-reinforced D.S.P. materials, however, are ductile. A fracture energy of 9 kJ/m^2 has been measured on samples of extruded D.S.P. materials with 4% (by volume) of 6 mm special polypropylene fibres.

R. J. Mangabhai (*Department of Chemistry, University of Salford, U.K.*). Has Dr Hjorth measured the mechanical properties of D.S.P. material at 8 °C and does the material set within 24 h? How long does the material have to be mixed for? With a high amount of Irgament Mighty is any retardation observed? Has Dr Hjorth done any rheological tests at 8 °C?

L. Hjorth. My group has no information on the properties of our materials at 8 °C. The mixing time varies with the composition of the D.S.P. material, and the type of mixer used. Intense mixing is required in order to obtain high quality products. High amounts of Irgament Mighty give retardation.

Sir Peter Hirsch, F.R.S. (*Department of Metallurgy and Science of Materials, University of Oxford*). Why is the tensile strength of Dr Hjorth's material so relatively low, i.e. one tenth of the compressive strength, if he has removed all the pores?

L. Hjorth. My group have changed the pore size distribution towards much finer pores, and reduced the total porosity, but have not removed all the pores.

G. K. Moir (*Blue Circle Industries p.l.c., Greenhithe, Kent, U.K.*). What are the curing conditions employed for a cast specimen such as the screw shown? In particular what is the temperature and duration of curing?

L. Hjorth. Accelerated curing at elevated temperatures (80 °C) is sometimes used in laboratory experiments to accelerate development programs. Industrial products, however, are normally cured at ordinary curing temperatures (20 °C) for a few days or weeks depending on the application.

S. A. Jefferis (*Civil Engineering Department, King's College London, U.K.*). In his last diagram Dr Hjorth showed a number of centrifugal pump impellers made from D.S.P. material. Can he say whether these can resist solutions that are normally detrimental to concrete, such as sulphate waters etc.?

L. Hjorth. Owing to the very fine pore structure of the D.S.P. materials the penetration of harmful ions, such as the sulphate ion, is very much restricted. The D.S.P. materials are still cement-based and they are therefore not inert to such ions. The service lifetime of the products in use, however, is very often sufficient to make them attractive.

R. Blundell (*Taylor Woodrow Research Laboratory, Southall, Middlesex, U.K.*). I was very interested to hear that the chloride ion diffusion coefficient is reduced by one order of magnitude compared with ordinary Portland cement concrete. Can Dr Hjorth confirm whether this can be achieved on site, as it appears to me that this material could have particular application to the repair of coastal and offshore structures where the diffusion of Cl^- ions is causing corrosion of embedded steel.

L. Hjorth. Although D.S.P. materials have already been used as described by Mr Blundell, we have not measured the rate of Cl^- penetration in such structures, mainly because the laboratory data are so convincing.

A. Kelly, F.R.S. (*University of Surrey, U.K.*). I really must congratulate Dr Hjorth upon a superb talk. It's so nice to see the ideas accepted of the advantage of low temperature forming of inorganics. Dr Hjorth mentioned fibres only in passing. What fibres have been added? What volume fractions of fibres?

L. Hjorth. Although the high strength D.S.P. materials demand high quality fibres and high fibre concentrations, my group have examined a variety of fibres and fibre concentrations. For instance, products have been made with 4% (by volume) of 6 mm special polypropylene fibres, and with 6% (by volume) of 6 mm steel fibres.

C. Defossé (*Dowell-Schlumberger Research and Development, Z. I. Molina La Chazotte BP* 90, *F*-42003, *Saint-Etienne Cedex, France*). Can Dr Hjorth, at the limit, make his paste pumpable? And if so, what are the best figures for plastic viscosity and yield value, on the assumption of a Bingham model?

L. Hjorth. The products can be made and have been made pumpable in practice. I have no figures for the rheological properties.

K. S. W. SING (*Department of Chemistry, Brunel University, Uxbridge, U.K.*). I should like to ask Dr Hjorth whether different grades of finely divided silica have been used to prepare high-density cement-based materials and if so what differences there are in their behaviour.

L. HJORTH. My group have been working with various types of microsilica from various ferro-silicon plants, and we have also worked with more expensive products, such as Aerosil. By making allowance for the differences of the products, it has been possible to make good D.S.P. materials from most of them.

F. MASSAZZA (*Italcementi Sp. A, Via Camozzi* 124, 24100 *Bergamo, Italy*). Does Dr Hjorth use special superplasticizers or high quantities of them, or both, to obtain such excellent results?

L. HJORTH. Most of the superplasticizers available on the market may be used. The amount of superplasticizers to be used depends on the type of D.S.P. material and the type of super-plasticizers, high dosages, however, are very often used.

J. BENSTED (*Blue Circle Industries p.l.c., Greenhithe, Kent, U.K.*). Has Dr Hjorth experienced any durability problems with his product, bearing in mind the low water/cement ratio and organic content, in wet enviroments?

L. HJORTH. D.S.P. materials have been stored in water for five years without showing any sign of deterioration.

Phil. Trans. R. Soc. Lond. A **310**, 175–190 (1983) [175]
Printed in Great Britain

Toughening of cement and other brittle solids with fibres

BY D. J. HANNANT, D. C. HUGHES† AND A. KELLY, F.R.S.
University of Surrey, Guildford, Surrey, GU2 5XH, U.K.

In common with other brittle solids, cements are toughened much more by the incorporation of fibres than by inclusions of other geometries. The largest energies required to break a specimen are found when multiple fracture of the specimen occurs before final failure. Theoretical models of a crack moving normal to a set of parallel fibres will be considered, to show that the crack spacing and first cracking strain should depend on the area of fibre–matrix interface per unit volume of composite. The first cracking strain is shown to increase for all fibre volumes provided that the fibre spacing is less than the critical flaw size according to the Griffith's equation. The theoretical models are compared with experiment and the practical difficulties of defining first cracking strain and interfacial area mentioned. The best practical means of assessing the resistance to failure of the composite is the work done per unit volume of the specimen in separating it into two distinct pieces. The maximum values of toughness attainable – some $10^6\,\mathrm{J\,m^{-3}}$ – can decrease with time under external weathering, owing to continuing hydration of the matrix and consequent increase in the critical volume fraction of fibres.

1. INTRODUCTION

(*a*) *Brittle solids*

Brittle solids are those that break without large amounts of plastic flow, so that the total work of fracture measured in, say, a controlled notch bend test of the Tattersall & Tappin (1966) type is less than about $1000\,\mathrm{J\,m^{-2}}$ *and* the strain to failure is no more than 1 % or so.

Small increases in toughness can be obtained by introducing pores (Cooper 1977; Coppola & Bradt 1973); by the introduction of rubber particles (Kunz-Douglass *et al.* 1980); by introduction of particles of a plastically deformable metal (Stett & Fulrath 1968; Krstic & Nicholson 1981) or by inclusion of a material that can undergo a stress induced phase transformation (Claussen 1978). Although these methods produce a measure of increase, the largest values of work of fracture that can be obtained fall very far short of those that can be achieved by the introduction of long fibres (see, for example, Phillips (1983) for a review).

These large increases are always accompanied by the appearance of the phenomenon of multiple fracture (Aveston *et al.* 1971). When multiple fracture occurs a series of more or less parallel cracks are formed in a body, which run in a direction approximately normal to the major principal axis of strain. The appearance of these cracks and their running completely across the body does not lead to failure in the sense of separation into two distinct pieces (although one component has certainly broken) and considerable extensions of the body are still possible in a direction normal to these cracks.

† Present address: Concrete Research Laboratory, Hatfield Polytechnic, College Lane, Hatfield, Herts. AL10 9AB, U.K.

It is important then to separate two quite distinct processes, which are:

(*a*) the (rapid) propagation of the first set of parallel cracks across the body, which starts at a more or less well defined strain, approximating that of the failure strain of the more brittle component tested separately and

(*b*) the subsequent process of absorption of energy in deforming the body to failure. This depends principally upon the properties of the high elongation component modified by the prior occurrence of (*a*).

Since the strains at which (*a*) and (*b*) occur are in general widely different and the total work of fracture is governed by (*b*), it is obvious that conventional fracture mechanics has no applicability in assessing the resistance to fracture of the body as a whole. Under these conditions the resistance to fracture is best, if crudely, assessed by quoting the work done on the body per unit volume in separating the specimen into two halves. This work per unit volume is the area under the stress–strain curve (Aveston & Kelly 1980). We give some examples for cement matrix materials in §§4, 5 below.

The rapid propagation of the first set of cracks, which traverse one of the phases, leaving the other essentially undamaged, may, however, be governed by some modified form of fracture mechanics, and this we discuss in §§2, 3. Before doing so, it is worth emphasizing that in the literature concerning fibre reinforced cements, a clear distinction is not always drawn between processes (*a*) and (*b*), which leads to confusion. If bend tests are done, process (*a*) may occur on a rising load–deflexion curve and a series of parallel cracks will traverse the tension surface of the specimen and then proceed to penetrate towards the interior. Irreversible work is done in this process and the use of the load–centre point deflexion curve, with the idea of the J integral, to obtain a measure of the potential energy available to crack the material (for example, by following the beautifully clear procedure of Begley & Landes 1972*a*, *b*) is in our opinion quite inadmissible.

The fracture of a monolithic piece of a brittle solid is usually rationalized nowadays by supposing it contains, either at the surface or in the interior, fissures or flaws of varying length, figure 1 (*a*). The propagation of the largest of these flaws, which is oriented normal to the applied stress, gives the breaking stress according to the formulation of Griffith (1920).

The addition of fibres to such a body may be assumed to affect the initial flaw size, limiting the length of flaws to the inter-fibre spacing, as in figure 1 (*b*); or leaving it unaltered, and hence with fibres traversing these cracks, figure 1 (*c*). If the process of introduction of fibres greatly alters the flaw distribution in the matrix then a discontinuous change in cracking strain of the matrix is expected to be observed on introducing fibres and there should be little connection between the cracking strain of the matrix without fibre and that observed when fibres are introduced. Alternatively, if the situation is that shown in figure 1 (*c*) a smooth increase in first cracking strain with introduction of fibres is expected.

Although there are many experiments that indicate an increase in first cracking strain with introduction of a high elongation phase into a brittle matrix or brittle laminate system, experiments with rather small volume fractions of the high elongation phase have usually been confined to cements (Aveston *et al.* 1974; Hughes 1983) and the nature of the initial flaw in cements needs a little discussion, which we give in §1 (*b*). Although there is rather smooth variation in first cracking strain with increase in volume fraction of the high elongation phase – which supports the model in figure 1 (*c*) – unfortunately, accurate measurements in tension of the cracking strain of the unreinforced matrix are exceedingly difficult.

Independently of whether the schema of figure 1 (*b*) or of figure 1 (*c*) apply in the initial stages of deformation, there is no doubt that the final stages can be described by a diagram like figure 1 (*d*). The condition for this to occur is that the high elongation phase is able to bear the total load.

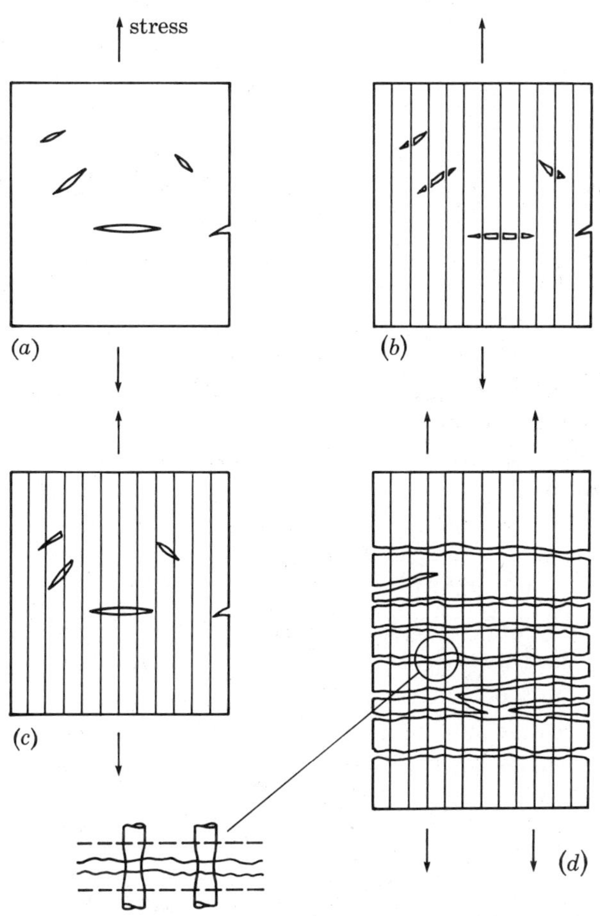

FIGURE 1. (*a*) Schematic representation of flaws in a brittle solid. (*b*) Schematic representation of the same solid, but now containing fibres, indicating that the initial flaw distribution is influenced by the introduction of fibres. (*c*) This is the same as (*a*), but indicates that the initial flaw distribution is not influenced by the introduction of fibres but that the fibres straddle the initial flaws, $s < 2c$ (see text). (*d*) Schematic representation of a cement paste that has undergone multiple fracture of the matrix (cement). The inset illustrates that the longitudinal stress in the fibres is not uniform.

The condition is

$$V_f \geqslant \sigma_{mu}/(\sigma_{fu}+\sigma_{mu}-\sigma'_f), \tag{1}$$

which, if both components are linearly elastic to the first cracking strain, can be written as

$$V_f \geqslant E_c \epsilon_{mu}/\sigma_{fu}, \tag{2}$$

where V is the volume fraction, E the elastic modulus, σ_u the failure stress, ϵ_u the failure strain, σ' the stress on the high elongation component when the other fails, and the subscripts m, f and c refer to the matrix, high elongation component and composite respectively.

If the volume fraction of high elongation phase is constant throughout the specimen and the interfacial conditions are also constant, then there is a lower limit to the spacing of the parallel cracks in figure 1 (*d*) given by

$$x = \sigma_{mu} \frac{V_m}{V_f} \frac{r}{2\tau} = \frac{\sigma_{mu} V_m}{\tau\beta}, \tag{3}$$

where β is the interfacial area per unit volume of composite (equal to $2V_f/r$ for aligned fibres of circular cross section) and τ is a shear stress.

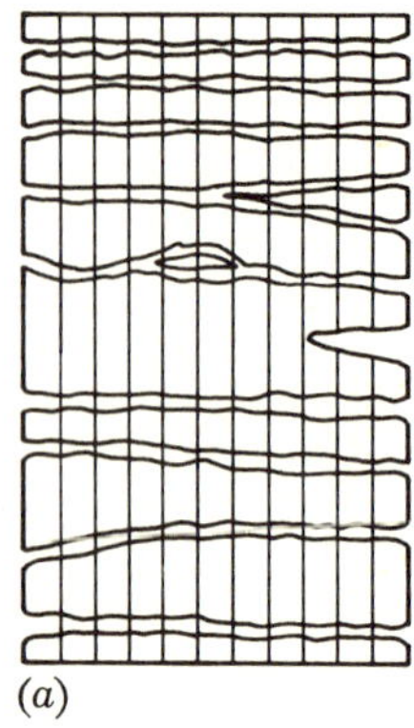

(*a*)

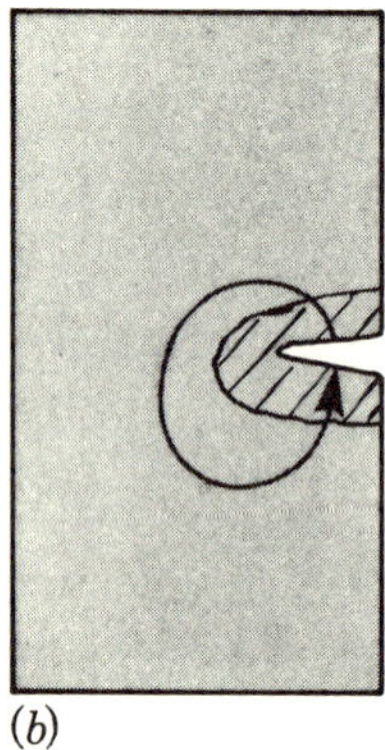

(*b*)

FIGURE 2. (*a*) A notched specimen containing fibres, which has undergone multiple fracture. (*b*) Model of a deformed notched specimen that is usually considered when one assesses fracture resistance with the *J*-integral.

If a specimen such as that in figure 1 (*d*) contained an artificially introduced notch or crack, which had been introduced by severing the matrix and high elongation phase, then its appearance would be as in figure 2 (*a*). Figure 2 (*a*) may be compared with figure 2 (*b*), which represents a specimen containing no high elongation phase. In bringing the specimen in figure 2 (*a*) to the state of multiple internal cracking with the consequent non-uniform elongation of the high elongation phase across the cracks, potential energy of the loading device has been used to fracture the matrix and cause *irreversible* sliding friction at the interface between the phases and remote from the notch, and so it is not all available to drive the notch further by fracture of the high elongation phase. The path for the J integral in figure 2 (*b*) cannot even be outlined on figure 2 (*a*).

(*b*) *Initial microstructure of cements*

There is currently, of course, as shown by this symposium, a good deal of interest in the elucidation of the microstructure of Portland cement paste. For the purpose of understanding the initial cracking strain and the work of fracture of the material we wish to emphasize two features. First, that the tensile stress–strain curve of the material is nonlinear, showing a decreasing slope after strains of less than 10^{-4}. A small permanent set may be observed on unloading from any stress, even one as low as 1 MPa. Part of this set at *room temperature* can be attributed to creep. These characteristics are shared with paper (see, for example, Seth & Page 1980) and cast iron (Haenny & Zambelli 1983). They are characteristic of a material that contains a large number of internal flaws and fissures that change their shape irreversibly under an applied load.

The breaking strength of such materials is found to be insensitive to the presence of 'small' notches cut into the surface because these notches are often 'shorter' or smaller in one dimension than the internal flaws already present in the material. Only if notches longer than a certain minimum size are cut does the strength depend upon their presence. For conventional cement paste this minimum size is about a millimetre (Higgins & Bailey 1976; Birchall *et al.* 1981).

The microstructure of a fibre reinforced cement must be thought of then as depicted in figure 3, where the matrix is of a more or less granular nature, consisting of partly hydrated cement grains bound together by a gel, with the whole containing voids that may be of size up to 1 mm, though more frequently of diameter 100 μm. There may, in addition, be long thin fissures of a crack-like

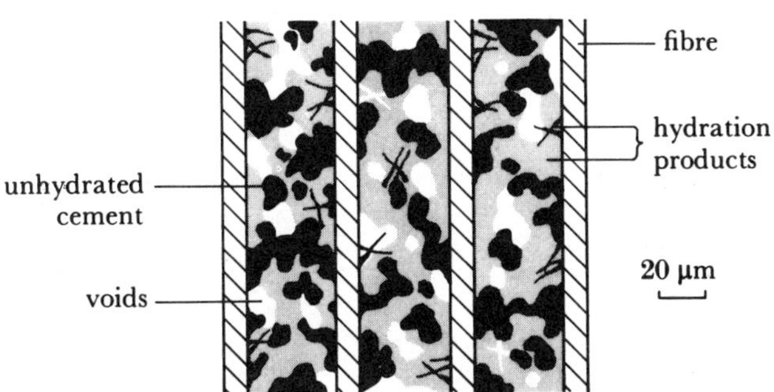

FIGURE 3. Schematic illustration of a cement matrix containing fibres. There are at least four phases present: fibres, unhydrated cement grains, hydration products and voids.

nature. The apparently smooth, well compacted surface of fibre reinforced cement cannot be regarded as really typical of the microstructure of the interior. It is very likely that there is no real adhesion between fibres and matrix and, as illustrated in figure 3, the interface will be characterized by more or less intermittent contact. Values derived for 'bond strengths' are then very uncertain (Laws 1982), and this is particularly so for polyalkene fibres in cement, which we shall be considering in §§3, 5. With polyalkene fibres, in addition, the values found for τ (cf. equation (3)) are extremely small, usually less than 0.5 MPa; an order of magnitude less than observed in other systems.

2. First cracking strain in the matrix

(a) *Without fibres*

If the matrix has a normally observed cracking strain, with no fibres present, of mean value ϵ_{mu}, then this can be interpreted to mean that it contains cracks of which the largest lying normal to the applied stress has a length $2c$, where $2c$ is given by the Griffith relation

$$\sigma/E = [g/E\pi(1-\nu^2)\,c]^{\frac{1}{2}} = \epsilon_{mu}, \tag{4}$$

where g replaces twice the surface energy in the original Griffith formulation, and a condition of plane strain is assumed.

This result can be derived (Friedel 1959; also see Hirth & Lothe (1968)) by considering an elastic equilibrium crack of which the length is $2c$ and the maximum opening B, so that B and c are related by the equation

$$\sigma/E = B/\pi(1-\nu^2)\,c. \tag{5}$$

If such a crack is to advance in a material that has a resistance to crack advance, g, then the applied stress acting through a distance equal to the maximum displacement, i.e. B, must do work at least equal to the resistance per unit area g. Hence, we must have

$$\sigma B \geqslant g. \tag{6}$$

By eliminating B between equations (5) and (6) we recover (4).

Thus at a critical crack opening B, the rate of release of strain energy exceeds the resistance to crack growth and the crack spreads in an unstable fashion.

If we apply equations (4) and (5) to Portland cement, taking $\epsilon_{mu} = 2 \times 10^{-4}$, $E = 30$ GPa, $g = 5$ J m^{-2} and $\nu = 0.25$, we find that the largest cracks present in the material have a length $2c \approx 2.8$ mm and a critical opening of 0.8 μm.

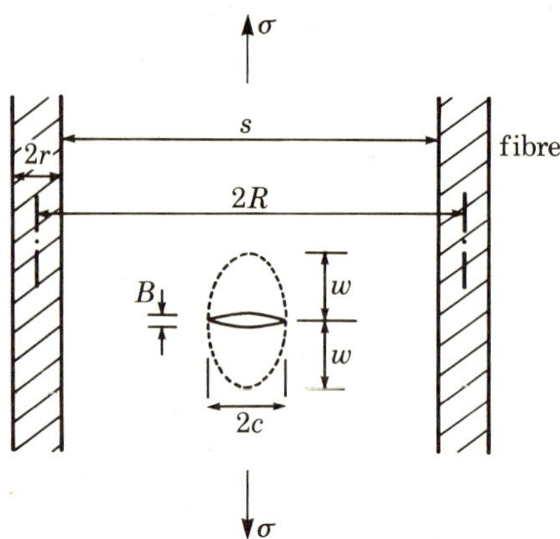

FIGURE 4. A crack under stress in a material with a fibre spacing much greater than the initial flaw size (see text for nomenclature).

(b) *Fibre spacing 's'* $\gg 2c$

When there are no fibres present, the matrix fails at an average strain given by equation (4). Assuming that the critical width for the flaw shown in figure 4 is B and that the matrix strain increases linearly from zero at the flaw to ϵ_{mu} at distance w, then the maximum relaxation distance w on each side of the crack may be found from

$$B = 2w \times \epsilon_{mu}/2 = w\epsilon_{mu}. \tag{7}$$

Substituting in equation (5) gives

$$w = \pi(1-\nu^2)\,c \simeq 3c. \tag{8}$$

Figure 4 shows the crack under stress, i.e. open, with width of opening B.

If fibres are introduced with a surface to surface spacing s, much greater than $2c$, see figure 4, the crack will advance unstably in the matrix when equation (4) is obeyed; with E equal to the modulus of the composite, before it encounters the fibres. When the unstable crack eventually meets the fibres, all the processes described by Aveston *et al.* (1971) are expected to occur provided that equation (1) is obeyed. If equation (1) is not obeyed and so the specimen breaks into two halves, there will be in principle a small increase in g if measured after total fracture (for example in a Tattersal & Tappin (1966) test), compared with the value of g for the matrix alone because of the need to deform the fibres to fracture.

An interesting prediction from equation (4) is that if the fibres added are of lower modulus than that of the matrix, then the cracking strain may actually be increased because E in equation (4) will be less than that of the matrix. It is important to determine for practical fibre reinforced systems over what range of volume fraction (V_f) equation (4) applies, i.e. whether s, the fibre spacing, is greater or less than $2c$.

(*c*) *Fibre spacing s in relation to 2c*

The surface to surface separation of the fibres is given by

$$s = 2(R-r) = 2r[(\alpha'/V_f)^{\frac{1}{2}} - 1], \tag{9}$$

where $\alpha' = \pi/2\sqrt{3} = 0.912$ for a hexagonal array of fibres and $\alpha' = \pi/4 = 0.785$ for a square array, R is the fibre separation (centre to centre) and r the fibre radius.

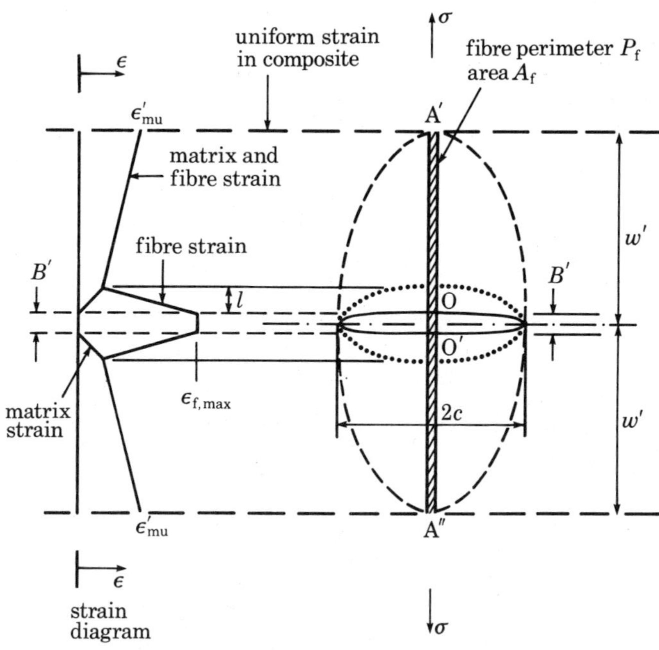

FIGURE 5. A possible zone of relaxation of strain around a crack under stress that is traversed by a single fibre. The left side of the figure illustrates the variation in strain within the fibre and the matrix along the line A′A″.

There will be at least one fibre within the critical crack length for the unreinforced matrix, if $2c$ equals s. So from equations (4) and (9)

$$2g/E_c\pi(1-\nu^2)\,\epsilon_{mu}^2 = 2r[(\alpha'/V_f)^{\frac{1}{2}} - 1]. \tag{10}$$

By taking values appropriate to a high strength Portland cement at w/c ratio = 0.2 and $E_c = E_m = 30\,\mathrm{GN\,m^{-2}}$, $g = 15\,\mathrm{J\,m^{-2}}$, $\nu = 0.25$, so $(1-\nu^2) \approx 1$, $\epsilon_{mu} = 4 \times 10^{-4}$ (0.04%) and a radius of fibre 50 μm, then equation (10) is satisfied for a volume fraction of fibre (V_f) greater than about 0.2%. The fibre spacing is then between 1 and 2 mm and so the argument is consistent with that given in § 1 (*b*).

For practical fibre cements containing asbestos fibres, glass fibre bundles, or polypropylene films at effective fibre volumes of 1.5%–10% the situation we must consider is therefore one in which $s < 2c$. Where steel fibres are concerned with r about 0.25 mm, effective fibre volumes of 3% would be required before $s \leqslant 2c$ and the lower fibre volumes *ca.* 1% used in practice would effectively rule out an increase in the applied strain before the cracks become unstable in the matrix.

For comparison, a hot pressed pyrex glass with $E = 70\,\mathrm{GN\,m^{-2}}$, $\epsilon_{mu} = 0.14\,\%$ and $g = 10\,\mathrm{J\,m^{-2}}$, containing carbon fibres of diameter 8 μm requires a V_f equal to about 1.7% before $s \leqslant 2c$. The volume fractions normally used exceed this figure (Phillips *et al.* 1972) and so we expect fibres to straddle the initial Griffith flaws. On the other hand, for a glass of the same properties, containing silicon carbide fibres of 140 μm diameter (Prewo & Brennan 1980), a volume fraction of greater than 40 % is needed before $s \leqslant 2c$, whereas one of the volume fractions they used was 35 %, fibres are therefore not expected to straddle the initial cracks.

(*d*) *Fibre spacing* $s \leqslant 2c$

When the fibre spacing is less than the Griffith flaw size we assume that the introduction of fibres does not by itself alter the length of the initial flaws in the matrix. We model the situation in figure 5, which for clarity shows the crack in the matrix as partially open. To expose the argument we suppose the crack to be traversed by a single fibre.

Imagine the strain in the specimen to be increased from zero with the 'crack' initially closed. At O and O′ the shear stresses between fibre and matrix will be rather large and if we assume that this shear stress can be described by a quantity τ, which is a constant independent of distance along OA′ and O′A″, then we expect the matrix to relax by sliding back over the fibres. The fact that the matrix has to slide back over the fibres means that the strain in the matrix will increase more rapidly along the lines OA′, O′A″, than it would have done in the absence of the fibres.

If the matrix slides back over a distance l a force $2\pi r\tau l$ is applied to a circular sectioned fibre. This produces a strain in the fibre of $\Delta\epsilon$, and, if we assume that the opening of the crack B' is solely due to this differential slip then

$$l = B'/\Delta\epsilon. \tag{11}$$

We also have

$$E_{\mathrm{f}}\Delta\epsilon = 2\pi r\tau l/\pi r^2 = 2\tau l/r. \tag{12}$$

So, by eliminating l, the stress in the fibre $\sigma_{\mathrm{f,\,max}}$ cannot be less than $E_{\mathrm{f}}\Delta\epsilon$ or

$$\sigma_{\mathrm{f,\,max}} \geqslant (2\tau B'E_{\mathrm{f}}/r)^{\frac{1}{2}}. \tag{13}$$

For $B' \approx 1\,\mu\mathrm{m}$, $\tau = 3\,\mathrm{MPa}$, $r = 5\,\mu\mathrm{m}$, $E_{\mathrm{f}} = 70\,\mathrm{GPa}$ (glass), $\sigma_{\mathrm{f,\,max}} = 290\,\mathrm{MPa}$, which is much greater than 28 MPa, the value of $E_{\mathrm{f}}\epsilon_{\mathrm{mu}}$ for $\epsilon_{\mathrm{mu}} = 4\times10^{-4}$.

We sketch the distribution of strain in fibre and matrix on the left side of figure 5. This sketch is made for a linear variation of strain with distance. We do not know that this variation is linear in reality of course, but it must be true that the fibre and matrix differ greatly in strain at the crack face and that the variation in each as one moves away from the crack is of opposite sense. Also, the fibre and matrix must attain the same strain before the edge of the relaxed region (within which the strain is less than the average strain of the specimen as a whole) is reached.

The fibres thus exert a closing force on the crack and reduce its opening. If there are N fibres crossing a crack, the total closing force, F, reducing the opening of the crack is

$$F \approx N\sigma_{\mathrm{f,\,max}}A_{\mathrm{f}}. \tag{14}$$

The quantity NA_{f} is proportional to the volume fraction of fibres. Hence we expect a continually increasing curve (from $V_{\mathrm{f}} = 0$) to relate fibre volume fraction with the matrix strain at which unstable crack propagation occurs, and it must be borne in mind that this is so whether or not the critical volume fraction given by equation (1) is exceeded.

3. Theory of the increased matrix failure strain

A theory of the increased failure strain of the matrix has been given by Aveston *et al.* (1971) and developed a little further by Kelly (1976). It applies when the strain energy released by a growing crack becomes independent of crack length. This applies when many fibres cross a crack and the quantity $\tau\beta$ is large. The theory will in general give a lower limit to the strain that must be exceeded for cracking to occur. Other attempts to explain the increased cracking strain have been given by Romualdi & Batson (1963), Tardiff (1973), Spurrier & Luxmore (1973, 1976).

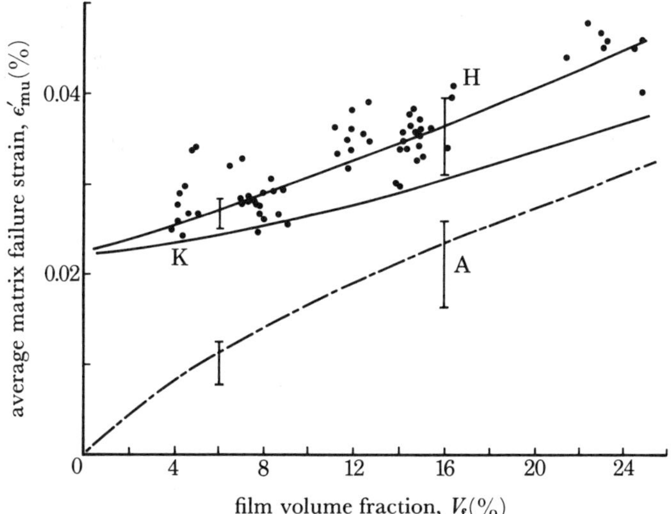

FIGURE 6. Relation between average matrix cracking strain and film volume fraction of a composite containing fibrillated polypropylene. (Effective fibre radius 23 μm, fibre modulus 7.7 GPa, matrix modulus 31.5 GPa, bond strength (τ) 0.5 MPa, matrix work of fracture 5 J m^{-2}, failure strain of unreinforced matrix 0.0222%, Griffith crack length 4.1 mm.) Upper and lower bars at V_f = 6% and 16% indicate the variation of failure strain predicted for values of τ between 0.2 MPa and 0.8 MPa. H, Hughes (1983); K, Korczynskyj *et al.* (1981); A, Aveston *et al.* (1971).

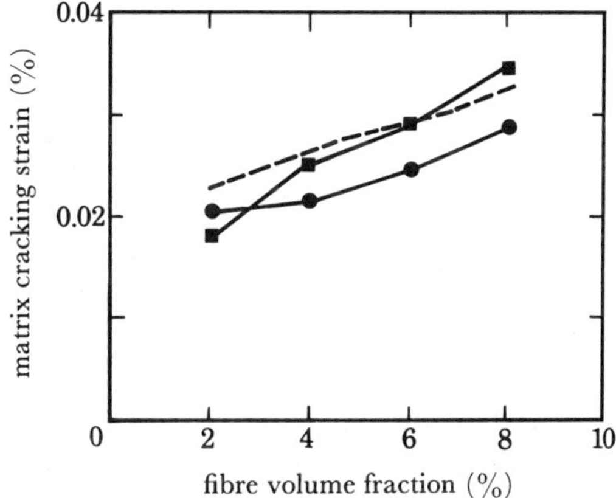

FIGURE 7. Theoretical and experimental values of enhanced matrix failure strain of cement that contains glass fibres randomly oriented in two dimensions. Experimental points after Ali *et al.* (1975). Theoretical curve from Hughes (1983). The assumed efficiency factor was 0.27. Squares represent data for specimens stored in water and dots are for specimens stored in air.

A computer calculation based on the idea of modelling a growing crack by assuming an initial shape for the crack and expanding it has been given by Korczynskyj *et al.* (1981). We set up a similar computer programme to theirs to see whether their theory applied to the results of Hughes (1983), who has made a very large number of measurements of the initial cracking strain for bars of cements containing continuous lengths of fibrillated polypropylene and polyethylene film. We believe that the assumptions of Korczynskyj *et al.* (1981) are not consistent with the occurrence of multiple fracture showing the experimental dependence on fibre volume fraction and the dimensions of the fibres and we therefore developed an alternative theory based on somewhat different assumptions. It is outlined in the Appendix, which is derived from Hughes (1983).

Theoretical predictions of the matrix failure strain are compared with experiment in figure 6, which contains Hughes's results. Figure 7 compares previously published results for glass fibre reinforced cement (Ali *et al.* 1975) with the predictions. Before commenting on these figures, it should be recognized that the Aveston *et al.* (1971) model (A.C.K. model) always gives a lower limit to the first cracking strain. This strain must be exceeded for cracking to occur and experimental results show the limit to be a valid one. However, as shown in §2(*d*), fibres arranged parallel to the applied stress and crossing a crack always hinder its growth and hence the presence of such fibres is always expected to increase the cracking strain of the matrix at all V_f when $s < 2c$, even if V_f is very small.

In figure 6, measurements of the matrix cracking strain from A.C.K. theory in the form

$$\epsilon'_{mu} = \left[\frac{6g_m \tau E_f V_f^2}{E_c E_m^2 r V_m}\right]^{\frac{1}{3}}, \tag{15}$$

the theory of Korczynskyj *et al.* (1981), and the theory from the Appendix are compared with experiment for specimens of fine grained ordinary Portland cement mortar containing between 4 % and 26 % (by volume) of continuous fibrillated polypropylene film. The parameters used to obtain the curves are given in the figure description. The difficulties of measuring the quantities, τ, r (effective radius of fibre), V_f and the moduli are given in detail in Hughes (1983) and will be published elsewhere. The value of the bond strength, τ, in equation (12), which applies to the initial fracture of the matrix, is particularly difficult to determine experimentally. Hence, the variation in calculated average matrix failure strain is also shown in figure 6 for bond strengths of 0.2 MPa and 0.8 MPa, which encompass the range of values calculated from equation (3) with data from tensile multiple cracking experiments. The computer model given in the Appendix gives good agreement with experiment. As emphasized in the Appendix, the quantity L_2 (see figure A1), which is a measure of the length of the relaxation zone on either side of the crack, is decreased with increase in β. It is this decrease that allows the model to be more consistent with the occurrence of multiple fracture than that of Korczynskyj *et al.* (1981). Quantitatively there is a problem, however, in that at 25 % fibre volume fraction we predict a length of relaxation of 3.5 mm on either side of a crack, whereas the crack spacing observed is a good deal smaller than this, typically less than 0.5 mm.

It should be borne in mind that a mere comparison with experiment of the theory of the variation of measured cracking strain with V_f is not a very stringent test of theory if that theory is scaled to pass through the observed unreinforced matrix cracking strain, as are both computer calculations. This is because the experimental scatter in cracking strain is 25 %–30 % and the increase in observed average value, even at volume fractions of 25 %, is less than a factor of two. Almost any theory that predicts an increase of ϵ_{mu} with V_f and passes through the value of ϵ_{mu} at $V_f = 0$ will be close to the experimental points.

The computer model given in the Appendix also accounts well for the experimental results on the increase of first cracking strain for glass reinforced cement reported by Ali (1975); see figure 7. In addition, the theory predicts, in accord with experiment, that the ultimate failure strain of the asbestos fibres in asbestos cement at $V_f = 0.05$ and 0.10, is reached before the crack becomes unstable. We therefore expect breaking of the specimen and first cracking of the matrix to occur simultaneously at a strain of 8.6×10^{-4} for $V_f = 0.05$, and at 1.2×10^{-3} for $V_f = 0.10$. These figures may not be directly compared with experiment because of the large volume of porosity in asbestos cement (Allen 1971) and because the fibres are not aligned. However, the prediction is in excellent accord with experiment.

4. Assessment of post-cracking toughness

Most of the energy absorbed in the complete fracture of fibre cements and concretes occurs after the matrix has cracked and can be calculated (Hibbert & Hannant 1982*a*). When fibre fracture occurs, the energy absorbed depends on whether the fibre volume fraction is greater or less than the critical value ($V_{f,\,crit}$) given by equation (2).

The area under the tensile stress–strain curve when expressed as energy per unit volume of material is in our opinion the most appropriate parameter to be used in design for energy-absorbing capability.† However, to compare results obtained at different laboratories and to calculate total energies required for failure, we recommend that all workers give the dimensions of their specimens so that both energy per unit volume and energy per unit area of final fracture surface can be calculated. The work done on the specimen in taking it through the complete stress–strain curve, when expressed as energy per unit volume, is given by

$$U = 0.5\,E_c\,\epsilon_{mu}^2 \quad \text{for} \quad V_f \leqslant V_{f,\,crit}, \tag{16}$$

$$U = 0.5\sigma_{fu}\epsilon_{fu}V_f + 0.159\,\alpha\,E_c\,\epsilon_{mu}^2 \tag{17}$$

for

$$V_f \geqslant V_{f,\,crit} \quad \text{and} \quad \alpha = E_m V_m / E_f V_f,$$

where, for fibres elastic to failure, the first term in equation (17) represents an upper bound to the strain energy required to strain the fibres to failure and the second term represents the minimum contribution from the multiple cracking of the matrix. The energy absorbed by multiple cracking is usually a relatively small proportion of the total energy.

5. Time dependent effects

Cements differ from most other brittle solids in that they show significant changes in properties with time under external weathering conditions although there is little change with indoor storage. Further, the fibre strengths may change with time in some composites. On economic grounds in practical composites, other than asbestos cement, the fibre volume is often kept to a minimum but care is taken to ensure that the critical fibre volume ($V_{f,\,crit}$) is exceeded so that multiple cracking occurs with a rising post-cracking tensile stress–strain curve. However, because of matrix changes under external weathering, and in particular an increase in E_m, the necessary critical fibre volume may increase substantially with time and if allowance is not made for this increase at the design and production stage, a composite that is initially tough and ductile with many cracks, may change its character to a brittle failure with a single crack.

† The energy absorbed before the component separates into two pieces is of importance when assessing the resistance of a structure to transient overloads such as random impact, wind loading, or explosion. It may not be related to criteria for acceptance under other conditions.

(a) *Hypothetical example*

To illustrate the effect that changes in the properties of the matrix can have on the energy absorbed to complete failure, we give an example within the range of practical composites. The properties of the matrix of a glass reinforced cement may alter with natural weathering to give a change in composite modulus (E_c) from about 22.5 GPa at 28 days to about 28.5 GPa after 5 years and 10 years (Building Research Establishment 1979). The matrix failure strain ϵ_{mu} at 28 days may be calculated from the bend-over point (i.e. where the stress–strain curve shows its first rapid change in slope) to be 4.2×10^{-4}, but it cannot be re-assessed after 10 years because of problems of testing. We therefore assume that it remains constant. Taking a value of fibre strength $\sigma_{fu} = 1000$ MPa and $E_f = 70$ GPa gives $V_{f,\,crit}$ from equation (2) equal to 0.95 % at 28 days and 1.2 % at 10 years. 1 % (by volume) of continuous aligned glass fibres exceeds $V_{f,\,crit}$ at 28 days and the energy, U, absorbed to complete failure will be 90 kJ m^{-3} from equation (17). After 10 years, however, V_f will be less than $V_{f,\,crit}$ and the energy absorbed will be only 2.5 kJ m^{-3} according to equation (16).

It is worth emphasizing that in this example both the glass content and strength are unchanged. The only changes that we postulated to have occurred are that the matrix has become stiffer and stronger and single fracture has replaced multiple fracture.

(b) *Experimental data for polypropylene films in cement*

We have obtained data on changes in elastic modulus and strain at the bend-over point (ϵ_{mu}) after 3 years natural weathering of composites containing 6 % (by volume) networks of fibrillated polypropylene film aligned with the stress. The data are:

$$28\text{ days},\quad E_c = 28\text{ GPa},\quad \epsilon_{mu} = 3 \times 10^{-4},\quad \sigma_{fu} = 280\text{ MPa};$$

$$3\text{ years},\quad E_c = 35.2\text{ GPa},\quad \epsilon_{mu} = 3.4 \times 10^{-4},\quad \sigma_{fu} = 270\text{ MPa}.$$

The critical fibre volume calculated from equation (2) increases from 3 % after 28 days to 4.4 % after 3 years. In this particular composite (Hannant & Keer 1983), where $E_f \approx 2.9$ GPa at strains greater than 1 % or so, the measured energy absorbed after 28 days was 790 kJ m^{-3} compared with a theoretical value (equation (17)) of 880 kJ m^{-3}. After 3 years the measured energy absorbed was 730 kJ m^{-3} compared with a theoretically predicted value of 890 kJ m^{-3}. Similar composites when stored under water for one year retain energies to failure in excess of 1 MJ m^{-3} (Hibbert & Hannant 1982*b*).

More recently produced polypropylene films have strengths of 600 MPa (Vittone *et al.* 1982) and these would require an increase in critical volume from 1.4 % at 28 days to 2 % at 3 years to maintain the high degree of toughness of the composite.

(c) *Glass fibre cement*

For glass reinforced cement, the following values have been published (Building Research Establishment 1979): 28 days, $E_c = 22.5$ GPa with a bend-over point at 9.5 MPa, which gives $\epsilon_{mu} = 4.22 \times 10^{-4}$, $\sigma_{fu} = 1000$ MPa; 10 years, $E_c = 28.5$ GPa, no accurate value for the bend-over point is given. If we assume no change in ϵ_{mu} and take σ_{fu} equal to 600 MPa after 10 years – for the recently developed fibre described by Proctor *et al.* (1982) – equation (2) gives an increase in $V_{f,\,crit}$ from 1 % at 28 days to 2 % at 10 years for aligned fibres.

Most glass reinforced cement is sprayed with short fibres in a random two-dimensional (2-D) orientation. To take account of the non-alignment, we multiply the value of σ_{fu} by 0.27 (the efficiency factor for stress, Oakley & Proctor (1975)) and find that a total fibre volume fraction of more than 3.5% at 28 days and more than 7.4% at 10 years is required to maintain ductility. Typical total fibre volumes are about 4% and it is known that the energy absorbed to failure, found from the area under the measured stress–strain curve (Majumdar & Laws, 1979), reduces from about 120 kJ m^{-3} at 28 days to less than 5 kJ m^{-3} at 5 years. This change may therefore be explained by the increase in critical fibre volume fraction with time.

Fibre concretes with less than the critical volume of short fibre at all ages, such as steel fibre concrete, may not suffer from such time dependent effects provided that fibre pull-out always remains the failure mechanism, although, the volume of specimen deformed to failure will be smaller if equation (2) is not obeyed.

Appendix

The model of crack growth developed by Korczynskyj *et al.* (1981) is based upon the assumption that the presence of fibres does not alter the length of the critical flaw or the size and shape of the relaxation zone around the flaw. Because the overall increase in length of the fibres within the zone (L_1 to L_2 in their figure 2) must be the same as that in the absence of the flaw, a proportion of the relaxed zone is at a constant strain, below that of the majority of the material. In §2 (*d*) it was suggested that the maximum opening of the flaw would be reduced in the presence of fibres, owing to the closing forces they exert. However, Korczynskyj *et al.* fail to predict any change in the flaw opening at fibre contents up to 25% (by volume).

A new computer model has been described (Hughes 1983) that does not assume a constant size of the relaxation zone. This model is summarized here. In common with Korczynskyj *et al.*, the relaxation zone around a flaw in the unreinforced matrix is assumed to be elliptical with major axis three times the flaw length and of the form $x^2/9+y^2 = c^2$, where $2c$ is the length of the flaw and the directions of x and y are shown on figure A 1. For the unreinforced matrix only the 'outer edge' of the ellipse is defined by a distance L_3 from the face of the flaw. The strain is assumed to increase linearly from zero at the face of the flaw to ϵ_m, the strain in the bulk of the material, at the edge of the ellipse, i.e. at a rate ϵ_m/L_3. The quantity L_3 does not define a distance from the crack in the *reinforced* matrix. This is the principal difference between the treatment here and that of Korczynskyj *et al.* (1981).

The presence of fibres is assumed not to alter the length of the critical flaw but alters the strain distribution and size of the relaxation zone as shown in figure A 1, which illustrates the situation immediately before unstable propagation of the flaw. Relative slip between fibre and matrix occurs for a distance L_1 from the face of the flaw. The strain in fibre and matrix is equal between L_1 and L_2 and is assumed to increase at a rate of ϵ'_{mu}/L_3.

The rate of change of strain in the matrix and fibre up to distances L_1 from the crack face is given by

$$\frac{d\epsilon_m}{dx} = \frac{2\tau V_f}{E_m V_m r} + \frac{\epsilon'_{mu}}{L_3}, \tag{A 1}$$

$$\frac{d\epsilon_f}{dx} = -\frac{2\tau}{E_f r} \tag{A 2}$$

respectively.

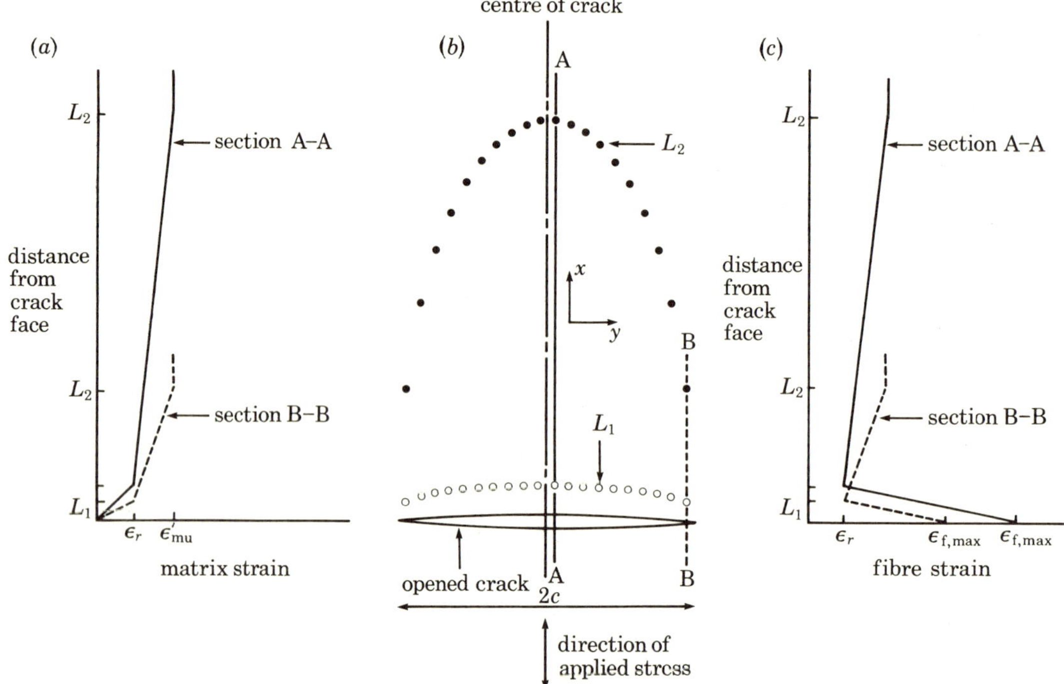

FIGURE A 1. Assumed distribution of strain in fibre and matrix along sections perpendicular to the crack face for a composite with a crack opened due to applied stress, fibre and matrix strains not to the same scale. L_2 (max.) $\equiv w'$ in figure 5 and also L_1 (max.) $\equiv l$ in figure 5. The strain, ϵ_r, at a distance L_1 is the same in fibre and matrix at a fixed value of y.

The strain in the matrix increases from zero to the average remote strain ϵ'_{mu}, figure A 1 (*a*). The strain in the fibre goes through a minimum at a distance L_1. The strain at a distance L_1 is $L_1 \mathrm{d}\epsilon_m/\mathrm{d}x$. The maximum fibre strain, $\epsilon_{f,\,max}$ is given thus:

$$\epsilon_{f,\,max} = L_1\left(\frac{\mathrm{d}\epsilon_f}{\mathrm{d}x}+\frac{\mathrm{d}\epsilon_m}{\mathrm{d}x}\right), \tag{A 3}$$

see figure A 1 (*c*). We also assume that

$$\epsilon_{f,\,max} = B''/L_1, \tag{A 4}$$

where B'' is the flaw opening at any distance 'y' from the centre of the flaw to the plane considered. Because of equations (A 1–A 4), B'' will vary with fibre volume fraction, which is different from the analysis of Korczynskyj *et al.* (1981).

At a distance L_1, the strains in fibre and matrix become equal to ϵ_r, such that

$$\epsilon_r = L_1 \mathrm{d}\epsilon_m/\mathrm{d}x. \tag{A 5}$$

The distances L_1, L_2 and L_3 are then given by

$$L_1 = \left(\frac{B''}{\mathrm{d}\epsilon_f/\mathrm{d}x+\mathrm{d}\epsilon_m/\mathrm{d}x}\right)^{\frac{1}{2}} \tag{A6}$$

from equations (A 3) and (A 4).

$$L_2 = L_1+(\epsilon'_{mu}-\epsilon_r)\,L_3/\epsilon'_{mu}. \tag{A 7}$$

L_2 defines the outer edge of the zone of relaxation when fibres are present, see figure A 1 (*a*).

$$L_3 = 3(c^2-y^2)^{\frac{1}{2}}. \tag{A 8}$$

The system is analysed by dividing the length of the flaw into segments of width, $\mathrm{d}y$, and unit thickness, where the length of the flaw, $2c$, is given by

$$c = g/E_{\mathrm{m}}\pi\epsilon_{\mathrm{mu}}^2, \tag{A 9}$$

i.e. assuming plane stress.

The first operation is to calculate the flaw opening such that the total extension of position L_2 is the same both inside and outside the relaxation zone, assuming the flaw to be fully closed in the unstrained condition.

The decrease in elastic strain energy (U_r) due to the presence of the flaw within each segment is given thus:

$$U_r = [\tfrac{1}{2}E_{\mathrm{c}}(\epsilon'_{\mathrm{mu}})^2 L_2 - \tfrac{1}{6}E_{\mathrm{m}}V_{\mathrm{m}}\epsilon_r^2 L_1 - \tfrac{1}{6}E_{\mathrm{f}}V_{\mathrm{f}}(\epsilon_{\mathrm{f,\,max}}^2 + \epsilon_{\mathrm{f,\,max}}\epsilon_r + \epsilon_r^2)\,L_1 - \tfrac{1}{2}E_{\mathrm{c}}(\epsilon'_{\mathrm{mu}})^2\,(L + L^3/3L_3^2 - L^2/L_3)] \times \delta y, \tag{A 10}$$

where $L = L_2 - L_1$.

The total decrease in elastic strain energy is obtained by numerically integrating equation (A 10) along the flaw, i.e. integrating y between the limits $-c \leqslant y \leqslant c$. This is repeated for a slightly larger value of c and then the rate of release of energy is obtained by dividing the difference between the two energies by the flaw extension.

Work is also done by friction, U_{f}, resulting from the relative displacement of fibre and matrix, so

$$U_{\mathrm{f}} = \frac{V_{\mathrm{f}}\tau\epsilon_{\mathrm{f,\,max}}L_1^2}{3r}\,\delta y. \tag{A 11}$$

A similar integration is performed as before, and to the subsequent rate of energy absorption is added the rate at which the matrix surfaces of the flaw absorb energy, i.e. gV_{m}.

The flaw will propagate catastrophically as soon as the rate of energy release becomes infinitesimally greater than the rate of absorption. In addition to the physical properties of fibre and matrix the computer requires an initial composite strain to examine for stability. Should the flaw be stable then the flaw is reset to its Griffith length and the process repeated with an increased strain.

The model has been shown to yield a satisfactory correlation with experimental data for polyalkene cements (see figure 6). It is predicted that the size of the relaxation zone is reduced by the presence of fibres, from a maximum 6.15 mm for the unreinforced matrix to 3.58 mm at 25 % (by volume) for the particular film considered. The maximum width of the critical flaw is similarly reduced from 1.43 μm to 1.04 μm respectively. These predictions are important since, if the model can be refined further such that (i) the length of the relaxation zone, either side of the flaw, is within the minimum crack spacing (equation (3)) and (ii) the model can allow for the numerous closely spaced flaws that are known to exist, it may be possible to fit a model, considering the growth of flaws, into the general multiple cracking theory of Aveston *et al.* (1971). This cannot be done for the model of Korczynskyj *et al.* (1981) because of the assumed constant size of the relaxation zone.

References

Ali, M. A., Majumdar, A. J. & Singh, B. 1975 Building Research Establishment, *Current Paper CP 94/75.*

Allen, H. G. 1971 *Composites* **2**, 98–103.

Aveston, J., Cooper, G. A. & Kelly, A. 1971 In *Conference Proceedings, National Physical Laboratory: The Properties Fibre Composites*, pp. 15–26. Guildford: IPC Science and Technology Press Ltd.

Aveston, J. & Kelly, A. 1980 *Phil. Trans. R. Soc. Lond.* A **294**, 519–534.

Aveston, J., Mercer, R. A. & Sillwood, J. M. 1974 In *Composites – Standards Testing and Design, N.P.L. Conference Proceedings*, pp. 93–103. Guildford: IPC Science and Technology Press Ltd.

Begley, J. A. & Landes, J. D. 1972*a* In *Fracture Toughness, Proceedings of the 1971 National Symposium on Fracture Mechanics, A.S.T.M. Special Technical Publication* no. 514, Part 2, pp. 1–20.
Begley, J. A. & Landes, J. D. 1972*b* In *Fracture Toughness. Proceedings of the 1971 National Symposium on Fracture Mechanics, A.S.T.M. Special Technical Publication* no. 514,Part 2, pp. 24–39.
Birchall, J. D., Howard, A. J. & Kendall, K. 1981 *Nature, Lond.* **289**, 388–389.
Building Research Establishment, U.K. 1979 *Paper* IP36, 79.
Claussen, N. 1978 *J. Am. ceram. Soc.* **61** 85–86.
Cooper, R. E. 1977 In *Fracture*, vol. 3, ICF4, Waterloo, Canada, pp. 809–818.
Coppola, J. A. & Bradt, R. C. 1973 *J. Am. ceram. Soc.* **56**, 392–393.
Friedel, J. 1959 In *Fracture propagation of cracks and work hardening* (ed. B. L. Averbach *et al.*). New York: Wiley.
Griffith, A. A. 1920 *Phil. Trans. R. Soc. Lond.* A **221**, 163–198.
Haenny, L. & Zambelli, G. 1983 *J. engng Fracture Mech.* (In the press.)
Hannant, D. J. & Keer, J. G. 1983 *Cem. Concr. Res.* **13**, 357–365.
Hibbert, A. P. & Hannant, D. J. 1982*a* *Composites* **13**, 105–111.
Hibbert, A. P. & Hannant, D. J. 1982*b* *Composites* **13**, 393–399.
Higgins, D. D. & Bailey, J. E. 1976 *J. Mater. Sci.* **11**, 1995–2003.
Hirth, J. P. & Lothe, J. 1968 *Theory of dislocations.* New York: McGraw-Hill.
Hughes, D. C. 1983 Ph.D. thesis, University of Surrey.
Kelly, A. 1976 In *Frontiers in materials science* (ed. L. E. Murr & C. Stein), pp. 335–364. New York: Marcel Dekker Inc.
Korczynskyj, S. J., Harris, S. J. & Morley, J. G. 1981 *J. Mater. Sci.* **16**, 1533–1547.
Krstic, V. V. & Nicholson, P. S. 1981 *J. Am. ceram. Soc.* **64**, 499.
Kunz-Douglass, S., Beaumont, P. W. R. & Ashby, M. F. 1980 *J. Mater. Sci.* **15**, 1109–1123.
Laws, V. 1982 *Composites* **13**, 145–151.
Majumdar, A. J. & Laws, V. 1979 *Composites* **10**, 17–27.
Oakley, D. R. & Proctor, B. A. 1975 In *Fibre Reinforced Cement and Concrete, RILEM Conf. Proceedings, London,* vol. 1, pp. 347–360. Lancaster: The Construction Press Ltd.
Phillips, D. C. 1983 In *Handbook of composites IV – technology* (ed. A. Kelly & S. T. Mileiko), ch. 7. Amsterdam: North Holland Publishing Co.
Phillips, D. C., Sambell, R. A. J. & Bowen, D. H. 1972 *J. Mater. Sci.* **7**, 1454–1464.
Prewo, K. M. & Brennan, J. J. 1980 *J. Mater. Sci.* **15**, 463–468.
Proctor, B. A., Oakley, D. R. & Litherland, K. L. 1982 *Composites* **13**, 173–179.
Romualdi, J. P. & Batson, G. B. 1963 *Proc. Am. Soc. civ. Engrs* **89**, 147–168.
Seth, R. S. & Page, D. H. 1980 In *J. tech. Ass. Pulp Paper Ind.* (Tappin), **63**, 99–102.
Spurrier, J. & Luxmore, A. R. 1973 *Fibre Sci. Technol.* **6**, 281–298.
Spurrier, J. & Luxmore, A. R. 1976 *Fibre Sci. Technol.* **9**, 225–236.
Stett, M. A. & Fulrath, R. M. 1968 *J. Am. ceram. Soc.* **51**, 599.
Tardiff, G. Jr. 1973 *Engng Fracture Mech.* **5**, 1–10.
Tattersall, H. G. & Tappin, G. 1966 *J. Mater. Sci.* **1**, 296–301.
Vittone, A., Camprincoli, P., Rossati, L. & Maltese, P. 1982 *Chimica Ind., Milano* **64** (9), 597.

Discussion

N. McN. Alford (*I.C.I. Runcorn, U.K.*). The beams that Dr Hannant showed, which exhibited multiple cracking, were tested in tension. If the beam is fractured under flexural centre point loading does multiple cracking still occur throughout the volume of the specimen even though there is a higher stress concentration at the load point?

D. J. Hannant. If a beam shows, anywhere on its surface, a strain in excess of ϵ'_{mu} we expect that part of the surface to show multiple cracking. Whether or not this strain, remote from the centre point (in three point bending), exceeds ϵ'_{mu} depends on the exact form of the rising load deflexion curve and on the dimensions of the specimen.

J. E. Bailey (*University of Surrey, U.K.*). The difficulty of obtaining values for the failure strain in tension for cement tests was mentioned. Dr Higgins did some careful tensile tests on well prepared and strain gauged specimens (w/c ratio 0.3). These specimens gave a failure strain *ca.* 0.05 %. Probably, in fibre composites, the matrix would be less perfect and this value can be regarded as an upper limit.

Phil. Trans. R. Soc. Lond. A **310**, 191–202 (1983)
Printed in Great Britain

Composite materials based on cement matrices

By A. J. Majumdar and V. Laws
Building Research Establishment, Building Research Station, Garston, Watford, Herts, U.K.

All materials produced from inorganic hydraulic cements are composites of one kind or another because of their multiphase nature. This paper briefly considers the relevance of composite principles in predicting the mechanical properties of hardened cement paste, mortar and concrete and then discusses recent developments towards enhancing these properties by the addition of polymers and more particularly fibres.

Fibres derived from glass, polypropylene and cellulose with metal wire are currently being used as cement reinforcements. Important properties of some of these practical fibre reinforced cement and concrete (F.R.C.) materials are discussed with particular reference to the replacement of asbestos products. Some examples of recent innovations in the manufacture of F.R.C. materials are given together with projections for their future use.

1. Introduction

Hydraulic cement paste, mortar and concrete can all be considered as composite materials because of their multiphase nature. Concrete, perhaps the most important construction material today, is most simply described as a two-phase system comprising aggregate particles embedded in hardened cement paste (H.C.P.) or alternatively coarse aggregates in mortar. To enhance the properties of cement mortar and concrete, various additions – polymers, fibres, special 'waste' materials and admixtures of different kinds – are now made and new industries have come into being to exploit these even more complex materials.

In this paper we discuss the composite nature of some of the cement products briefly before describing the current developments in one of the newer areas, that of fibre reinforced cement and concrete (F.R.C.). Unless otherwise stated the cement referred to in this paper is ordinary Portland cement (O.P.C.).

2. Hardened cement paste, mortar and concrete

Several investigations have shown (Hansen 1968) that the elastic constants of concrete can be predicted from the elastic constants and volume fractions of the constituents by using two-phase models and mixture rules, and that such an approach can be extended to H.C.P. and mortar. However, the agreement between predicted and observed values is not very close when simple models (Reuss or Voigt) are used. Several alternative models have been proposed and Hobbs (1973) has shown that the equation for elastic constants derived by Hashin & Shtrikman (1963) from variational principles gives very good agreement between theory and experiments. These models, however, rely on assumptions that are not strictly valid and an alternative approach has been suggested, namely fitting exponential and power law curves (see, for example, Kotsovos & Newman 1979).

Properties can be altered very substantially by the addition of polymers, and several different types of polymer modified H.C.P., mortar and concrete have been developed in the last two decades (Manson 1976). Of these polymer impregnated concrete (P.I.C.) is the most interesting and also the most difficult to manufacture on a large scale. The polymer greatly enhances the durability of the concrete and can improve its mechanical properties severalfold. In general a glassy polymer such as polymethyl methacrylate (PMMA) produces a solid that is linearly elastic almost up to the point of failure, but the stress–strain behaviour can be modified by introducing plasticizing co-polymers. Although the role of the polymer in P.I.C. is understood in a qualitative way, satisfactory analytical models have not been demonstrated yet. Current ideas about the nature of the polymer in the pore system of H.C.P., the extent of chemical reaction between the polymer and H.C.P., and interfacial effects remain largely speculative.

Molten sulphur is a far simpler impregnant and Feldman & Beaudoin (1977) have recently made a detailed study of the properties of H.C.P. and autoclaved cement/silica mixtures impregnated with sulphur. Uniform and nearly total impregnation was possible in some cases and the elastic constants of the composites could be predicted with reasonable accuracy from the mixture rule (Reuss) by using the properties of the constituents at zero porosity.

Constitutive relations have been derived for several other properties of concrete (Hansen 1968), for example shrinkage and thermal expansion, creep, thermal and electrical conductivity. However, it has not been possible to describe the strength of concrete (or of H.C.P.) in this way since it fails in a progressive manner, owing to its inherently complex structure, and failure criteria are not well established.

3. Fibre reinforced cement and concrete

(a) *Constitutive relations*

It is usual to regard F.R.C. as consisting of two phases, fibres and matrix, and relations have been derived to predict the tensile and bending properties of the composite in terms of the properties and volume fractions of the two components, and of their interactions in terms of the bond between them.

(i) *Elastic properties.* In the elastic region simple mixture rules have been developed but usually any improvement in elastic modulus is small. There is some evidence to suggest that fibres can modify the failure criterion and therefore the failure strain of the matrix, but the comparatively large increases predicted by the elastic theory of Romualdi & Batson (1963) are not seen in practice. Kelly (1974) attributes this to debonding and loss of elastic continuity at the interface. The non-elastic theory of Aveston *et al.* (1971) predicts a smaller increase in matrix failure strain. Experimentally, strict confirmation of the prediction is difficult to obtain. However, if the addition of fibres leads to a more reliable matrix-cracking strain, this in itself would have considerable practical significance.

(ii) *Stress–strain curve and composite strength in tension.* The effect of fibres that has received most attention is that applying after the matrix has failed, and has been defined by Aveston *et al.* (1971). The A.C.K. theory supposes that, after the first matrix crack has formed, there are sufficient fibres present to support the load. Predicted stress–strain curves in tension deduced by Aveston *et al.* (1974) are shown in figure 1 for cement reinforced with continuous aligned carbon fibres. The agreement between the two sets of curves (predicted and experimental) is good, and generally the A.C.K. theory appears to describe the stress–strain curves of F.R.C. materials containing more than the critical volume of fibres exceptionally well.

When the reinforcement is in the form of a two-dimensional random array of short fibres, one approach is to replace the response to stress of the continuous aligned fibres of the A.C.K. model by that of a mat of short fibres held across a crack (Laws *et al.* 1971). The response to stress of the fibrous mat defines the shape of the tensile stress–strain curve after cracking of the matrix has ended, the strength of the composite, and the curve after the composite has failed, as illustrated in figure 2. The response consists of two parts, namely the contributions of the fibres that 'hold', that is the elastic response, and that of the fibres that 'slip', the frictional response. After the maximum stress has been reached there is still a mainly frictional component (a post-failure stress capacity) as fibres that have slipped are pulled out of the matrix and one crack opens up. The strength of the composite, σ_c, is the maximum stress that the fibrous mat can support and is given by $\sigma_c = \eta\sigma_f V_f$, where σ_f and V_f are the ultimate tensile strength and volume fraction of the fibres and η is an efficiency factor that depends on the orientation of the fibres, their length, l, the strength of the bond opposing slipping, τ_s, and that of the frictional bond, τ_d, operating over the fibres that are slipping. The term ϵ_f is the fibre failure strain and $_{\mathrm{C}}$ i*l*s the critical fibre length.

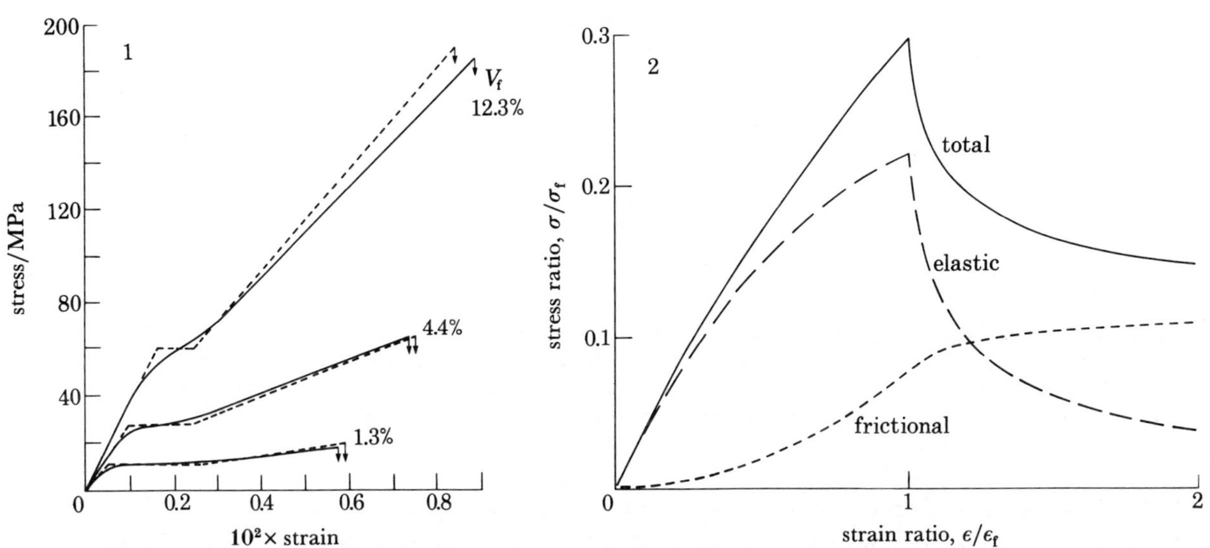

FIGURE 1. Tensile stress–strain diagrams for continuous carbon-fibre reinforced cement. Full lines are the experimental curves, broken lines are the curves predicted by the A.C.K. theory (Aveston *et al.* 1974).

FIGURE 2. The stress–strain response of a fibrous mat, for $l/l_c = 2$ and $\tau_d/\tau_s = 1$. The broken line refers to the proportion of fibres that hold (the elastic response); and the broken line with shorter strokes to those that slip (the frictional response). The full line is the total response.

(iii) *Bending properties and m.o.r./u.t.s. ratio*. Commonly, even linear materials are stronger in bending than in tension (the size effect) but the ratio of the modulus of rupture (m.o.r.) to the ultimate tensile strength (u.t.s.) is usually much lower than 2. F.R.C. materials are markedly non-linear in tension and this non-linearity alone leads to m.o.r./u.t.s. ratios up to a theoretical maximum of 3. Aveston *et al.* (1974) have developed constitutive relations for the m.o.r. and the ratio m.o.r./u.t.s., based on the A.C.K. model. In practice, prediction with these relations is problematical because of the uncertainty about the values of the variables involved, and bending curves are commonly computed directly from measured tensile and measured or assumed compressive stress–strain curves (Allen 1971). However, the computed values often fall short of those measured directly and in particular the apparent strain at failure in bending is often higher than that predicted. This discrepancy can be accounted for, at least in part, by the post-failure stress capacity in tension and the 'size effect' (Laws & Walton 1978).

The bending test is widely employed in materials testing to obtain an indirect measurement of the tensile strength; for fibre cement composites, correction factors (m.o.r./u.t.s. ratios) must be applied and this leads to uncertainties. There is also a problem in determining the limit of proportionality (l.o.p.) in bending. Experiments suggest that for F.R.C. materials the l.o.p. in bending is considerably higher than that in tension. Part of this increase may be attributed to a size effect; but part might also be the result of the increased uncertainty in locating the first deviation from linearity on the bending curve.

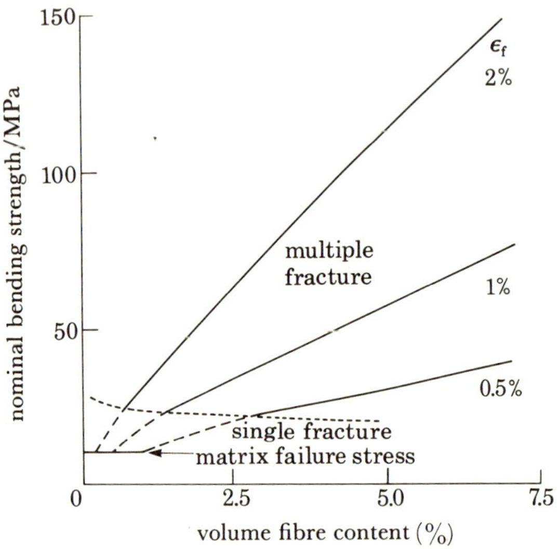

FIGURE 3. Predicted bending strength as a function of fibre content, for Young moduli E_f 76 GPa and E_m 25 GPa, ϵ_m 0.04 %, and increasing values of ϵ_f. The full lines refer to the A.C.K. model composites (V_f above critical V_f for reinforcement in tension); the broken lines refer to composites with V_f below critical V_f in tension and above critical V_f in bending. The broken line with shorter strokes separates the multiple fracture and the single fracture conditions.

There is a further effect of the bending–tensile relation that appears to have been overlooked. It is generally agreed that for F.R.C. materials that fail after multiple cracking, the u.t.s. is given by $\eta\sigma_f V_f$ and there is no contribution from the matrix. But experimental results, for example on carbon fibre cement (Briggs *et al.* 1974) suggest that a mixture rule might apply for the bending strength. The relation between the m.o.r. and the u.t.s. for F.R.C. offers an explanation. At low fibre volume fractions the tensile stress–strain curve after matrix failure is flat and the m.o.r./u.t.s. ratio is high; at high fibre volumes the tensile curve approaches linearity and the m.o.r./u.t.s. ratio is low. When these factors are applied to a linear u.t.s. against V_f relation, the resulting m.o.r. against V_f relation is biased upwards at the low fibre volume end. It then has the appearance of a two-phase mixture rule. Figure 3 shows the apparent bending strength against V_f relation computed from the A.C.K. model. Extrapolation of these solid lines to zero fibre content suggests a positive intercept, i.e. a matrix contribution. The figure also shows the implied effect in the single fracture region between critical fibre volume fraction for reinforcement in tension and the critical fibre volume fraction for reinforcement in bending. In this region the composite fails in tension at the matrix failure strain ϵ_m; a mixture rule should apply although the effect of V_f would be small. The work of Swamy & Mangat (1974), however, suggests a much bigger effect and one that depends on the fibre–matrix bond strength and the fibre aspect ratio. It is difficult to visualize, as implied in the theory of Swamy & Mangat, how the fibres can all slip at the matrix failure strain and it is much more attractive to suppose that the post-failure 'tail' during fibre pull-out, which leads to a critical V_f for reinforcement in bending (Hannant 1978), might be responsible for the observed results.

(iv) *Work to break.* The energy required to extend and to break the composite follows directly from the tensile stress–strain curve. The energy to break an aligned continuous fibre composite described by the A.C.K. model is the area under the load–extension curve up to fibre failure. A large part of this energy is the elastic strain energy of the fibres.

When the fibres are short, by using the fibrous mat model (figure 2) and neglecting the constraint of the matrix in reducing the average fibre strain at failure, it can be shown that for $l \geqslant l_C$ the contribution W of the fibres to the composite strain energy per unit area up to fibre failure strain is

$$W = \frac{3}{16}\left\{\left(1-\frac{5}{9}\frac{l_C}{l}\right)+\frac{5}{18}\frac{l_C}{l}\frac{\tau_d}{\tau_s}\right\}\sigma_f\epsilon_f V_f L, \tag{1}$$

where the first term is the elastic contribution and the second is that arising from frictional slip. The term L is the length of the specimen. The relative size of the two terms follows from equation (1); for example if $\tau_d = \tau_s$, the ratio of frictional to elastic energy is $\frac{5}{26}$ when $l = 2l_C$ and increases to $\frac{5}{8}$ for $l = l_C$.

The total energy per unit area, W', needed to pull out the fibres that do not break is, for $l \geqslant l_C$,

$$W' = \frac{2}{\pi}\left(\frac{1}{12}\frac{\tau_d}{\tau_s}\frac{l_C^2}{l}\sigma_f V_f\right) = \frac{1}{6\pi}\frac{\tau_d r^2}{l}\left(\frac{\sigma_f}{\tau_s}\right)^3 V_f. \tag{2}$$

For a model composite containing 5% (by volume) of fibres of strength 1200 MPa, length 30 mm and critical fibre length 15 mm, the energy to break calculated from equation (2) is approximately 24 kJ m^{-2} when $\tau_d = \tau_s$. The strain energy to maximum stress ratio calculated from equation (1) is 0.17 kJ m^{-2} per mm of sample strained, or 10 kJ m^{-2} if a specimen of length 50 mm is uniformly strained. Of this, less than 2 kJ m^{-2} is the frictional component.

The contribution of the fibres to the total energy to break a specimen and pull out the fibres is (approximately) the sum of the irrecoverable part of the strain energy W and the pull-out energy W'. If the elastic strain energy is largely recovered as one crack opens and the others close, the total energy to break for the model above is approximately 25 kJ m^{-2}, which is within the range of measured impact strengths of young glass reinforced cement (G.R.C.) samples.

There has been some discussion of the relative importance of the bond strength and the fibre tensile strength in determining the energy needed to break a fibrous composite. The ratio σ_f/τ_s is directly related to the critical fibre length l_C and determines the proportions of fibres that slip and that hold. From equation (1) it follows that a high elastic energy contribution from the fibre requires a low l_C together with a high fibre strength, hence a strong bond; and the maximum possible value of this is $\frac{3}{16}\,\sigma_f\epsilon_f V_f L$. The maximum work to pull-out, however, requires that all the fibres slip and that τ_d is as high as possible, i.e. that $\tau_d = \tau_s$ since $\tau_d \leqslant \tau_s$. The relation between pull-out energy, fibre strength and bond strengths τ_s and τ_d is illustrated in figure 4. The discontinuities in the curves mark the conditions where all fibres slip, further increase in σ_f then has no effect on pull-out energy; but an improvement in τ_d has a marked effect.

(*b*) *Asbestos cement replacement*

The properties of different types of F.R.C. materials have been reviewed recently by the ACI Committee 544 (1982). There is much general interest in the performance of the relatively cheap melt-extracted stainless steel fibres that have been introduced successfully in the refractory-castable industry. Some experiments have been done at B.R.E. recently with different types of melt-extracted stainless fibres. Sheets of reinforced O.P.C. mortar, 10 mm thick and

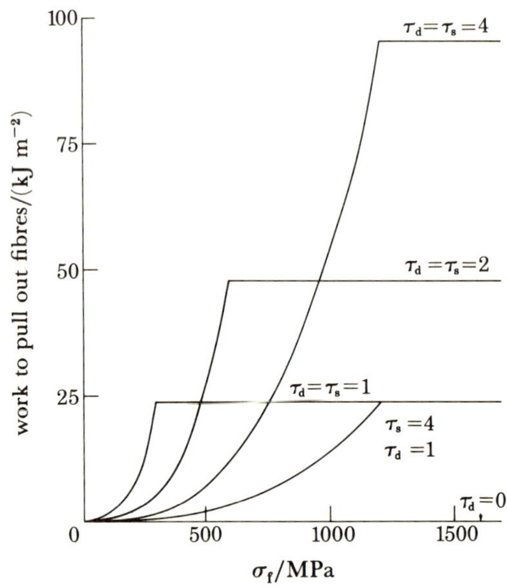

FIGURE 4. Work to pull-out fibres, calculated from equation (2), as a function of fibre strength σ_f and bond strengths τ_d and τ_s (for V_f 0.05; l 30 mm and fibre radius 0.1 mm). Values for τ_d and τ_s are quoted in MPa.

incorporating up to 6% (by volume) of fibre, were made by the spray-up method. Strength values, m.o.r. of 15–20 MPa and u.t.s. of 6–8 MPa are not particularly impressive although impact resistance (Izod) values 11–15 kJ m^{-2} are good. No significant changes in the properties of these materials have been observed up to 2 years under natural weathering conditions. The composite boards were, however, difficult to manufacture and handle, and were very hard to cut and saw.

Many of the new fibre cement composites are serious contenders for replacing different types of asbestos products traditionally used in the construction industry. In the development of asbestos-free fibre cement composites, in roofing applications for example, two distinct trends are discernible. Traditional asbestos cement manufacturers are choosing the alternative fibre mainly in terms of the ease with which it can be handled by the existing manufacturing techniques within the industry, particularly the Hatschek process. Developers outside the asbestos cement industry have more freedom in this respect and entirely new production methods are being attempted, for instance, to incorporate polyolefin-film fibre nets (Bijen & Geurts 1980). The cost of the fibre and the properties of the new composite, including its durability, are other major considerations in fibre selection.

Wells (1982) has recently described the properties of some of the asbestos cement alternatives produced on the Hatschek machine (table 1). The mechanical properties of the composites, for example carbon fibre cement, can be improved considerably by increasing the fibre volume fraction, but present fibre costs (Harper 1982) are such that the alternatives may then be less attractive than other materials, say PVC-coated steel. From the point of view of costs, cellulose fibres (for example, wood pulp) have obvious advantages and asbestos-free insulation boards containing large amounts of cellulose are already in use (Harper 1982). These autoclaved cellulose reinforced alternatives are dimensionally stable as the matrix is not H.C.P., but consists mainly of crystalline calcium silicate hydrates with very little, if any, free lime. In the cement matrix, the long term viability of wood pulp or other vegetable fibres (Cook 1980) remains questionable.

Table 1. Properties of asbestos cement substitutes manufactured on the Hatschek machine (After Wells 1982)

product	fibre content (% by mass)	dry Charpy impact strength kJ m^{-2}	modulus of rupture/MPa dry	wet	wet aged at 50 °C	combustibility (BS 476, Part 4)
asbestos cement	10	2–4	30	25	—	pass
alternative fibre composite based on a.r. † glass (CemFIL)	3.5	4.6	16	11.5–12.3	9–10 (200 days)	pass
a.r. mineral wool	5	3.5–4.5	18	17.2	8.2 (84 days)	pass
polypropylene monofilament	0.5–2	6	16–17.5	16–17.5	—	fail, 70 °C temperature rise, 600 s flaming
carbon	1.0–2.0	2.5–4.5	18–22	17–19	16.0 (122 days)	pass
refined cellulose	3.5	1.5–2.5	15–18	13–15	11.0 (84 days)	pass
cellulose‡	9	3.9	19.5	12.2	—	fail, 65 °C temperature rise, Class 1 surface spread of flame

† Alkali-resistant. ‡ Autoclaved calcium silicate matrix.

Several other fibres not included in table 1 are currently receiving attention in industrial laboratories as asbestos substitutes and judging from the patent literature it would appear that fibres derived from polyvinyl alcohol are of particular interest. Much research has also been done on glass reinforced cement (G.R.C.) made by the spray-up process, and on cement composites containing continuous nets of very fine fibrillated polyolefin film fibre. The latter material produced by a lay-up technique (Hannant & Zonsfeld 1980) is very tough and claimed to be strong enough to satisfy the requirements for corrugated roofing. But as with other alternatives containing organic fibres the combustibility test of BS 476 may pose a particular problem (Wells 1982).

Alkali-resistant glass fibre CemFIL (trade mark of Fibreglass (U.K.) Ltd) has been the cornerstone of the G.R.C. industry during the last decade. By using a high throughput spray-up method G.R.C. sheets can now be manufactured on a nearly continuous basis with speeds of 10 m/min or more, which is comparable to the production rate of asbestos cement by the Hatschek process. The properties of spray-dewatered G.R.C. of this type containing *ca.* 4% (by volume) of CemFIL fibre in different environments for up to 10 years have been the subject of detailed studies (B.R.E. 1979; Proctor 1980), and very long-term (more than 100 years) projections of some of these properties have been made from the results of accelerated ageing tests and natural weathering in various parts of the world (Proctor 1982). These studies have shown that under wet and natural weathering conditions the strength and impact resistance of

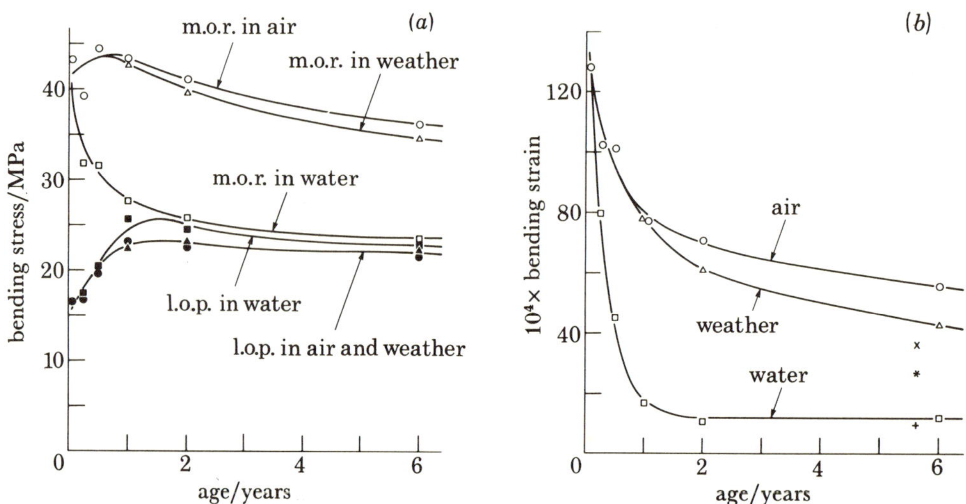

FIGURE 5. Bending properties of polymer modified G.R.C. in different environments as a function of age. (*a*) Modulus of rupture and limit of proportionality: ○●, in air; □■, in water; △▲, in weather, (*b*) Average strain in the outermost fibres at maximum load; ○, in air; □, in water; △, in weather. Other symbols: x, G.R.C. control in air; +, G.R.C. control in water; *, G.R.C. control in weather.

CemFIL G.R.C. made from O.P.C. are significantly reduced with time and this factor must be taken into account in designing with G.R.C. In relatively dry air, however, G.R.C. properties remain more or less stable with time and lightweight G.R.Cs (West *et al.* 1980) incorporating, for instance, pulverised fuel ash cenospheres have considerable potential as substitutes for asbestos-containing materials in internal applications such as insulation and fire protection.

The long-term properties of CemFIL G.R.C. in wet environments are improved when non-Portland cements such as high alumina and supersulphated cements or modified O.P.C. such as pozzament are used (Majumdar 1980; Majumdar & Singh 1982). In natural weather the addition of certain types of polymer dispersions may also be beneficial. This option has been exploited in the development of a type of G.R.C. (Forton), where the reinforcement is provided by borosilicate E – glass fibres and 15% (by volume) of an acrylic polymer emulsion is added to the O.P.C. mortar matrix (Jacobs 1982). The flexural properties of a G.R.C. sheet (10 mm thick) produced by the spray-dewatering method at B.R.E., incorporating 4% (by volume) of CemFIL fibres, an O.P.C. matrix, and containing 11.5% (by volume) styrene-acrylic co-polymer solids, kept in three different environments up to 6 years are shown in figure 5*a*, *b*. It is interesting to note that in natural weather (B.R.E. site) the polymer modified G.R.C. has retained a significant proportion of its initial strength after six years although its strain capacity has been reduced. The reasons for the much improved performance of the polymer modified materials in natural weather, in comparison with that under water, are not clear.

The emergence of CemFIL 2 fibre, which is more alkali-resistant than CemFIL (Proctor 1982), may lead to the use of G.R.C. as a replacement for asbestos cement. Smith (1982) has recently reported that asbestos cement products such as corrugated roofing, pipes, guttering etc. can be made in CemFIL 2 G.R.C. by using standard asbestos cement production methods. He claims that by using 3 % (by mass) fibre and a modified O.P.C. matrix, a corrugated roofing sheet in G.R.C. can be manufactured that will pass the existing British Standard for the corresponding asbestos cement product and will incur only a small cost penalty. The properties of two composite boards, made at B.R.E., each containing *ca.* 5 % (by mass) CemFIL 2 fibres, but very different matrices, are given in table 2. For both boards weathering results up to 2 years are highly encouraging, particularly in respect of ultimate failure strain and impact resistance. Accelerated ageing test results indicate that for the granulated slag–cement composite, the properties should not change very much with time unless factors such as carbonation become important. For CemFIL 2–O.P.C. mortar composites some reduction in properties with time would be expected.

TABLE 2. PROPERTIES OF CEMENT COMPOSITES CONTAINING *ca.* 5 % CemFIL 2 GLASS FIBRES

	70 % O.P.C. 30 % sand (by mass)			70 % granulated slag 30 % O.P.C. (by mass)		
property	28 days	natural weathering at B.R.E. for 2 years	wet aged at 50 °C for 180 days	28 days	natural weathering at B.R.E. for 2 years	wet aged at 50 °C for 50 days
modulus of rupture/MPa	38	36	24	35	36	37
limit of proportionality in bending/MPa	12	11	12	11	15	11
ultimate tensile strength/MPa	15	13	9	13	15	13
ultimate tensile failure strain in microstrain	11000	8000	870	9800	9000	9500
tensile Young modulus/GPa	30	33	33	22	23	33
Izod impact strength/(kJ m^{-2})	21	20	8	21	19	20

While seeking replacements for asbestos cement it should be borne in mind that the replacement must meet the performance specifications for particular applications. National and International Standards on asbestos cement products may require important amendments or additions, or both, if they are to be used as the basis for specifying similar products in other F.R.C. materials.

(*c*) *Future prospects*

The past decade has seen numerous applications of F.R.C. materials in construction (Hannant 1978; ACI 1982). Many of the current uses of steel fibre reinforced concrete, for example in pavements and overlays, hydraulic structures and refractory concrete, will no doubt continue but in future more use is likely to be made of fibre reinforcement in selected zones of the component or structure. In precast components, fibre concentrations could be increased greatly by prepacking the moulds with fibres first (Lankard & Lease 1982). More research will be done to improve the effectiveness of fibres by placing them in a favourable disposition. The glass fibre reinforced spun concrete pipe (Farahar 1978) in which continuous strands of glass fibres in large concentrations are positioned in the inner and outer layers only, provides a very good example. Magnetic alignment of short steel wires in concrete components has also been achieved (Skarendahl 1980). For economic reasons, general use of fibres in reinforced concrete structures may be limited to special cases such as concrete reactor vessels and blast or seismic resistant buildings. A recent use of steel fibre reinforced concrete in the construction of hemispherical domes by using an inflated membrane process may also be a pointer for the future.

Fibre reinforced cement sheets are energy efficient in comparison with similar products in metal or plastics and their predominant constituent, i.e. cement, is non-combustible. The advantages of F.R.C. materials have been exploited with great enthusiasm by the G.R.C. industry in uses such as cladding, permanent formwork and repair of old buildings and structures. The reduction of strength of the first generation alkali-resistant glass fibres in cement has limited the range of G.R.C. applications (B.R.E. 1978), but with the emergence of CemFIL 2 and the promise of an even more alkali-resistant fibre (Proctor 1982), the future prospect of G.R.C. as a material has improved considerably. One may see more widespread use of the material in low-cost housing or as encasements for steel reinforced concrete beams and columns.

In many of these thin sheet applications several other fibres are suitable also. Carbon and aramid fibres produce excellent cement composites, and if their cost can be reduced such composites could be developed further for structural use. Composites reinforced with nets of fibrillated polyolefin film fibre may become commercially available in the near future and their usefulness may be extended further by the introduction of the high-modulus fibre varieties (Ward 1980). In future we may see more widespread use of mixtures of fibres in cement reinforcement.

Much more attention will be paid in future to matrix formulations. The addition of pozzoanas, fillers, admixtures, polymers etc. to an O.P.C. matrix will probably increase, and so might the use of non-Portland cements. It would be interesting to see if fine fibres can be introduced in strong low-porosity cements, such as the recently developed 'macro-defect-free' (MDF) cement (Birchall *et al.* 1981), to provide extra toughness. The principles behind fibre reinforcement of cement and concrete are reasonably well understood now but there are still uncertainties in some areas such as the importance of the fibre–matrix bond and the best method to measure bond strength or the efficiency factors that designers might need to use to predict composite properties. Further work should be aimed at resolving these issues.

The work described has been done as part of the research programme of the Building Research Establishment of the Department of the Environment and this paper is published by permission of the Director.

References

Allen, H. G. 1971 *J. comp. Mater.* **5**, 194–207.

American Concrete Institute Committee no. 544 1982 *Concr. Int.* (*Design Construction*) **4**, 9–30.

Aveston, J., Cooper, G. A. & Kelly, A. 1971 In *Conf. proc. N.P.L.: The properties of fibre composites*, pp. 15–26. Guildford: I.P.C. Science and Technology Press, Ltd.

Aveston, J., Mercer, R. A. & Silwood, J. M. 1974 In *Conf. proc. N.P.L.: Composites – standards testing and design*, pp. 93–103. Guildford: I.P.C. Science and Technology Press, Ltd.

Bijen, J. & Geurts, E. 1980 In *Concrete Society Symposium Fibrous Concrete*, pp. 194–202. Lancaster: Construction Press Ltd.

Birchall, J. D., Howard, A. J. & Kendall, K. 1981 *Nature, Lond.* **289**, 388–390.

Briggs, A., Bowen, D. H. & Kollek, J. 1974 In *Proceedings 2nd International Carbon Fibres Conference*, pp. 114–121. London: The Plastics Institute.

Building Research Establishment 1978 Glass fibre reinforced cement. *B.R.E. Digest* no. 216. London: H.M.S.O.

Building Research Establishment 1979 Properties of glass reinforced cement: ten year results. *B.R.E. Information Paper IP* 36/79.

Cook, D. J. 1980 In *Concrete Society Symposium Fibrous Concrete*, pp. 99–114. Lancaster: Construction Press Ltd.

Farahar, R. M. 1978 *Precast Concr.* **9**, 559–564.

Feldman, R. F. & Beaudoin, J. J. 1977 *Cem. Concr. Res.* **7**, 19–30.

Hannant, D. J. 1978 *Fibre cements and fibre concretes.* New York and Chichester: John Wiley and Sons.

Hannant, D. J. & Zonsfeld, J. J. 1980 *Phil. Trans. R. Soc. Lond.* A **294**, 591–597.

Hansen, T. C. 1968 In *The structure of concrete and its behaviour under load* (ed. A. E. Brooks & K. Newman), pp. 16–23. London: Cement and Concrete Association.

Harper, S. 1982 *Composites* **13**, 123–128.

Hashin, Z. & Shtrikman, S. 1963 *J. Mech. Phys. Solids* **11**, 127–140.

Hobbs, D. W. 1973 *Cement and Concrete Association technical report* no. 42-484.

Jacobs, M. J. N. 1982 In *International congress proceedings: GRC in the* 80*s*, pp. 31–49. Gerrards Cross, U.K.: Glass Fibre Reinforced Cement Association.

Kelly, A. 1974 In *Conf. proc. N.P.L.: Composites – standards, testing and design*, pp. 9–16. Guildford: I.P.C. Science and Technology Press Ltd.

Kotsovos, M. D. & Newman, J. B. 1979 *Mag. Concr. Res.* **31**, 77–90.

Lankard, D. R. & Lease, D. H. 1982 *J. Am. ceram. Soc. Bull.* **61**, 728–732.

Laws, V., Ali, M. A. & Nurse, R. W. B. 1971 In *Conf. proc. N.P.L.: The properties of fibre composites*, pp. 29–30. Guildford: I.P.C. Science and Technology Press Ltd.

Laws, V. & Walton, P. L. 1978 In *RILEM Symposium: Testing and test methods of fibre cement composites*, pp. 429–438. Lancaster: Construction Press, Ltd.

Majumdar, A. J. 1980 In *Concrete Society Symposium Fibrous Concrete*, pp. 48–68. Lancaster: Construction Press, Ltd.

Majumdar, A. J. & Singh, B. 1982 Non-Portland cement glass fibre reinforced cement. *B.R.E. Information Paper IP* 7/82.

Manson, J. A. 1976 *Mater. Sci. Engng* **25**, 41–52.

Proctor, B. 1980 In *Concrete Society Symposium Fibrous Concrete*, pp. 69–86. Lancaster: Construction Press Ltd.

Proctor, B. 1982 In *International Congress Proceedings: GRC in the 80s* pp. 50–67. Gerrards Cross, U.K.: Glass Fibre Reinforced Cement Association.

Romualdi, P. J. & Batson, G. B. 1963 *Proc. Am. Soc. civ. Engrs* **89**, 147–168.

Skarendahl, A. 1980 In *Concrete Society Symposium Fibrous Concrete*, pp. 115–127. Lancaster: Construction Press, Ltd.

Smith, J. W. 1982 *Composites* **13**, 161–163.

Swamy, R. N. & Mangat, P. S. 1974 *Cem. Concr. Res.* **4**, 313–325.

Ward, I. M. 1980 *Phil. Trans. R. Soc. Lond.* A **294**, 473–482.

Wells, R. A. 1982 *Composites* **13**, 169–172.

West, J. M., Majumdar, A. J. & de Vekey, R. C. 1980 *Composites* **11**, 19–24.

Discussion

J. E. Bailey (*Department of Metallurgy and Materials Technology, University of Surrey, Guildford, U.K.*). In view of the need for research on the durability of glass fibre reinforced cement building products, is it true to say that sufficient attention has been paid to the effect of the environment on pipes made out of these materials for the sewerage and drainage applications briefly touched upon in this paper?

A. J. Majumdar. The pipe described by Farahar (1978) has undergone a rigorous testing programme that has included accelerated ageing, and the acceptance of the pipe in various countries is based on the favourable results obtained. Existing information on the long-term properties of G.R.C. flat sheets is not strictly applicable to the G.R.C. layers in the pipe because the latter are constructed in a very different way and have large concentrations of highly oriented continuous glass-fibre rovings.

The long-term performance of the pipe in drains and sewers may depend primarily on the type of cement and the quality of concrete in these pipes.

D. B. Downey (*ARC Concrete Ltd., U.K.*). It is only fair to respond to Professor Bailey's question and to put his mind to rest in respect of sewer failures. For glass reinforced concrete pipe there has been a programme of accelerated testing underway for the past ten years, which clearly

demonstrates strength retention. This evidence is supported by successful performance under service conditions for over seven years, and currently more than 100 km of Slimelinepipe is functioning effectively within the nation's sewer asset. Similar test work in South Africa and Japan is also supportive. The available data have been studied by a specialist group brought together by the British Standards Institution, whose conclusion is published in DD76 stating that it is reasonable to conclude that no strength loss will be experienced with these pipes in service.

In regard to acid attack, it is well known that H_2S generated in conditions of low flow, shallow gradient and hot climate can attack concrete. Glass reinforced concrete pipes are only used in circumstances where other more traditional concrete pipes are acceptable. It is therefore reasonable to conclude that sewer failures are unlikely in the short, medium or long term.

C. D. Pomeroy (*Cement and Concrete Association, Slough, U.K.*). ARC pipes have been thoroughly studied and the method of manufacture is not a randomly reinforced fibre mat, but the pipes are a true sandwich composite. There is no need to worry about their use in sewers. The inner skin is a very dense and impermeable layer and this will provide extra protection against chemical attack.

J. Bensted (*Blue Circle Industries p.l.c., Research Division, Greenhithe, U.K.*). Has Dr Majumdar investigated the influence of cement alkali content upon the long term behaviour of G.R.C. composites?

A. J. Majumdar. We have not studied the effect of the alkali content (Na_2O+K_2O) of cement on the long-term properties of G.R.C. in a systematic manner. However, over the years several different batches of O.P.C. having different alkali contents have been used in our work. G.R.C. composites made from these cements have shown very similar trends in long-term durability.

I may mention in this connection that Proctor *et al.* (1982) have recently published results that indicate that the range and type of alkali content commonly encountered in rapid hardening Portland cements in the U.K. has little effect on the degree of strength reduction suffered by CemFIL fibres when they are placed in the cement matrix.

Reference

Proctor, B. A., Oakley, D. R. & Litherland, K. L. 1982 *Composites* **13**, 173–179.

Phil. Trans. R. Soc. Lond. A **310**, 203–207 (1983) [203]
Printed in Great Britain

Concluding assessment of future development

By G. M. Idorn
G. M. Idorn Consult ApS, 14B Tovesvej, 2850 Nærum, Denmark

The contributions to the conference are reviewed with the aim of identifying their potential influence on the development of hydraulic cement science and technology in the 1990s.

The high-technology–low-volume innovations presented are emphasized because they point towards future efforts in this direction.

Most of the contributions commented upon are related to the basic materials, the processing characteristics and the performance behaviour of concrete and cement paste. The availability of fly ash and slag is referred to because it is encouraging more determined research.

The need for more concern about R. & D. systems as means to transform research into reliable technology, accessible to the practising engineer and labour forces, is stressed.

Introduction

The presentations and discussions confirm that the Royal Society has chosen an appropriate time to arrange this symposium on the development of science and technology of hydraulic cements in the 1990s. It is interesting to learn what science and research currently have to offer for improvements in technology because they are urgently needed now, and will be needed during the forthcoming decades for rehabilitation work, the reduction of investment costs in building and construction and the improvement of labour motivations. Improvements are also needed for social development in the 'third world', where there are many instances that demonstrate that conventional hydraulic cement technology does not meet the demand for the production of reliable, cost-effective buildings and construction works. One may also, from the contents of the contributions, sense concern about the restructuring of science and research itself, an issue that was discussed in the U.S.A. in the *1977–1980 National Materials Advisory Board study* (N.M.A.B. 1980) referred to by J. P. Skalny.

It is apparent that for there to be significant impact in our societies from the realization of the technology opportunities presented requires larger and more coherent R. & D. projects than those done now, and in particular clearer strategy and short term initiatives regarding effective R. & D. planning and management, appropriate science and engineering education and co-operation between universities and industry. The science education aspects are particularly important, because it is tempting for engineers and cement-based industries to rely upon university involvement in trouble-shooting and research during the present economic constraints, rather than acquiring the competence for front-line competitiveness by themselves. Therefore, the creation of a large number of brilliant, outspoken students, to question much of what has been said at this and similar preceding meetings, is needed to make substantial improvements in the performance of the 8–10 Gt of cement products referred to. Consequently, further elucidation of the educational issues seems commendable.

I wish to stress that it is the engineers who face the challenge to gain public confidence in the reliability of the research, when it is to be applied as a technological development.

High-technology–low-volume innovations

The present symposium is surely the first in the history of cement and concrete to devote a section of its programme to high-technology–low-volume innovations with hydraulic cements. This is definitely a pointer towards the future, and amidst the present, general recession and low morale in many heavy industries it is encouraging and admirable that two high-technology innovations presented have emerged from Great Britain. They both have their origins in co-operation between university and public research on the fundamental side, and large corporations outside the cement and concrete community on the technical–commercial side. They are namely the alkali-resistant glass-fibre reinforced composites, presented by A. J. Majumdar, and the MDF cements, presented by J. D. Birchall *et al.* Both inventions are the result of a willingness to take financial risks in a period with severely constricted markets.

It is interesting that the result of an equally daring example of innovative spirit, the silica fume based D.S.P. products, is presented from one of the smallest countries present, by L. Hjorth.

Each of these materials is about ready for rigorous feasibility testing in the markets. At present they demonstrate commendably the determined coherence of systematic R. & D. as described in educational works by the European Industrial Research Management Association and by the Industrial Research Institute in the U.S.A.

Low-technology–high-volume development

The fundamental urgency of cement based technology development can adequately be categorized as demands on improved basic materials, processing, and concrete performance behaviour.

Basic materials

The technology of cement-making has been impressive and has kept ahead of the cement-using technologies throughout its development, since the British Aspdin patent no. 5022 of 1824. The presentations by W. A. Gutteridge & C. D. Pomeroy, G. R. Long. G. W. Groves, G. K. Moir, and, in some ways, also that by J. P. Skalny & G. Frohnsdorff are those among the contributions that are most oriented towards the cement itself, though inadvertently the authors support the judgement that radical changes of cement raw materials, composition and manufacture technology are not waiting 'around the corner'. This seems justified because the cement industries have seen considerable innovations during the last 20–30 years, and today with preheater and precalciner kilns, alkali by-pass, computerized on-line process monitoring, etc., are well prepared to meet even considerable technology improvements among the cement uses.

There are, as yet, unanswered challenges in the potential development of the utilization of about 500 million tons of fly ash and blast-furnace slag per year, not merely for greater use of these materials, but also for more effective access to the hydraulic energy inherent in them. Incidentally, when Professor Birchall says that the uses of cement 'have been restricted by the low tensile strength and fracture toughness of hardened cement paste' – this is true – but with some modifications. The total cement use in the world was about 25 million tons in 1920, is about 850 million tons now, and the demands, seen socially, are such that during the 1990s we will approach 2000 million tons. Consequently, the restrictions referred to have been and are of limited effect.

Processing

It appears from all the contributions that the processing of concrete and cement products has not been the sole aim for the presented research, nor has it been used as the most important criterion for predictions regarding application of the research results in technology in engineering and industrial practice. It seems that the scientific community considers further clarification of the fundamental nature of the hydration reactions and their basic structure formation phenomena desirable, before the consensus aims at more practical development.

Despite apparently unsettled discrepancies about some of the fundamental aspects of cement hydration, the eight papers by R. G. Ottewill, D. D. Double, G. W. Groves, M. Regourd, P. Pratt, J. E. Bailey & C. J. Hampson, G. K. Moir and L. J. Parrott all fit into the 'concrete processing' category. They illuminate, expressed in simple terms, 'how the stuff gets stiff and gains strength and longevity'.

M. Regourd made some very interesting remarks about the modifications of surfaces of clinker minerals and slag, which take place as soon as they are mixed with water; millions of workmen and engineers are awaiting more guidance on how these discoveries can lead to improved monitoring of workability in concrete making. L. J. Parrott's observations are compatible and emphasize that pore structure is formed early during hydration. This points to the need for supplementary theoretical studies of the effects of vibration and chemical admixtures. R. H. Ottewill's contribution in its entirety is dealing with workability, if considered from an engineering viewpoint, though few civil engineers would recognize the characteristics in the presentation without further guidance. Other contributions also probably include observations of similar significance. However, the emphasis seems to be on the hydration chemistry and on its related morphology and submicroscopic structure in cement paste.

J. E. Bailey & C. J. Hampson are exploring the 'yellow area' on the hydration map of aluminate phases. I do hope that the alumina-ferrite phases will also be covered in these studies, as there are many unsolved problems in using cements with high–low C_3A against low–high $C_2(A, F)$ content. I do wonder if the alumina phases really come out entirely as fibrous $C\bar{S}A$ hydrates. The optical microscope does not reveal that, and recent work (Komarneni *et al.* 1982) suggests Al-ion exchange capability in alumina-hydrate phases. Thus, in concrete, there may be many different pathways for Al during performance conditions.

P. Pratt is one of the few authors to discuss fly ash. That is very timely because so much new empirical research is being published that nobody can really assess the fundamental nature of fly ash hydration – its kinetics and thermodynamics – and without this knowledge money will be wasted and failures will occur.

The advance of the basic knowledge since physico-chemical methodology was forcefully brought to bear, about 40 years ago, is formidable, has admirably and steadily been updated by the development of modern instrumentation, and will be used in any future development towards greater manufacturing effectiveness and improved quality assurance for cement uses. Notwithstanding this advance, there are two issues requiring more attention if the further basic studies of hydration are to become useful for progress in technology, these are rheology and curing. Fibre reinforced composites, MDF cement and D.S.P.-type materials do represent conscious application of basic rheology. Concrete, in contrast, is largely made with, at best, crude modelling of its behaviour during mixing, transport and placement, if any model is used at all. Development in this area is urgently needed, even in current concrete practice. Concrete has, during the last 10–20 years, been made a reacting system, significantly sensitive to variations in materials and to changes in operation conditions. And skilled workmen must frequently perform wonders when monitoring workability with but little background of explanatory research.

Most current hydration research deals with systems reacting at room temperatures. Until 30–40 years ago concrete did experience slow temperature rises to 30–40 °C during the curing phase, and a slow decline to ambient temperatures in the temperate climatic regions, so that the room temperature modelling was justified. Much contemporary concrete reaches 80 °C or higher temperatures within 24 h of casting and rapid subsequent cooling down to ambient temperatures is enforced. No concrete or cement products are cured at constant room temperature. Only test and research specimens are processed this way. There are numerous studies to show that at curing temperatures above 50–60 °C, the permeability and porosity of cement paste become considerably increased, and ultimate strength is decreased. And there are numerous incidents to justify the view that much current and forthcoming repair work on concrete buildings and structures is due to high curing temperature deterioration of the concrete.

Extrapolation of the room temperature projection of the kinetics of cement hydration to deal with the temperature variation of the hydration process is so much more rewarding, because in some countries fully computerized systems for monitoring curing are in operation for design and construction work. Danish concreting guidelines for winter and highway construction specifications are examples.

Recent research (see, for example, Regourd 1980; Bamforth 1980; Wesche & Schubert 1982; Roy & Idorn 1982) has demonstrated the significant beneficial effects of heat on the hydration of fly ash and slag, accompanied by its moderation of the temperature development during Portland cement hydration in blended cements. Altogether, the state of affairs and the progress of research on cement hydration presented at this symposium are promising to supply a broader basis for monitoring development of concrete processing in engineering and industry within a few years, if made to include adequate progress of knowledge on rheology of fresh cement paste mortars and concrete, and realistic approaches to the kinetics of hydration. This will leave 'the room temperature syndrome' behind as a historically justified relic of the pre-industrial era.

Concrete performance behaviour

There are four contributions in this category to consider, namely those by R. W. Davidge, K. Kendall, A. J. Howard & J. D. Birchall, and D. J. Hannant *et al.* Until the superplasticizers became available a few years ago, preceded by many years of theoretical research by P. A. Rehbinder and workers, 'ceramic' strength of cement paste was a vision beyond insurmountable obstacles, except for very special laboratory procedures, as referred to by L. Hjorth. The introductory contributions to this conference do show remarkable achievements for overcoming the barriers. The field of knowledge presented by R. W. Davidge, K. Kendall, and also by D. J. Hannant *et al.*, comes much closer to integrated applications than before. Once again, slag, fly ash and silica fume do contribute to increased density, and to higher proportions of ultrafine pores in cement paste. Recent work (Short & Page 1982) is of interest in this connection.

Considering fibrous composites, it is a striking development in technology that, owing to lack of timber, about 50% of all river barges and boats in China are now made of ferrocement. They serve a very large part of the national transportation system for 1100 million inhabitants. Do not assume that urgent needs in developing nations will wait for accomplishments in basic research if overwhelming social problems are at stake!

In some respects these four papers bridge the initial distinction of practice between high and low technology development, because they do illuminate that the properties that are now believed to sell high-technology cement-based products are desirable, though less easily attainable; also for voluminous concrete bodies and elements. *Density, porosity, toughness and deformability are all, in small and large bodies, decisive parameters of strength and stress transfer capability.*

Research should not overlook that the same parameters are important for the longevity of the materials, provided that basic materials' compositions and processing have assured a long term convergence of the reacting systems towards thermodynamic equilibrium under the given exposure conditions.

Conclusions

The outlined prospects for the hydraulic cements and cement uses are probably expected to be what industry and engineering, at the present especially, will want as a basis for their technology development in the 1990s. If research planning with the objective to satisfy such requirements is in conflict with strong desires in the science oriented community for research to make progress in basic knowledge *per se*, then a period with turbulent, dialectic development of the research is predictable. Maybe, out of necessity, this is the most fruitful way to progress in the near future. But whichever course we see emerging, the proceedings of this symposium are likely to become an influential source of information on essential fragments of the entirety. Thus, the Royal Society's approach should be truly rewarding to everybody concerned.

References

Bamforth, P. B. 1980 *Proc. Inst. civ. Engrs.* **69**, 777–800.

Komarneni, S., Roy, D. M. & Roy, R. 1982 *Cem. Concr. Res.* **12**, 773–780.

Regourd, M. 1980 In *Proc. 7th Int. Congr. Chem. Cem., Paris*, vol. 2, III. 105–III. 111. Paris: Éditions Septima.

Roy, D. M. & Idorn, G. M. 1982 *J. Am. Concr. Inst.*. **79**, 444–457.

Short, N. R. & Page, C. L. 1982 *Silicates Industriel* **10**, 237–240.

The Status of Cement and Concrete R. & D. in the United States 1980 Paper NMAB–361, (117 pages). The National Research Council.

Wesche, K. & Schubert, P. 1982 In *Proc. Symp. Nat. Mater. Res. Soc. Annual Meeting, Boston*, pp. 225–232.